Study and Solutions Guide for

TRIGONOMETRY

SECOND EDITION

Larson/Hostetler

Dianna L. Zook

The Pennsylvania State University
The Behrend College

D. C. Heath and Company

Lexington, Massachusetts   Toronto

Published simultaneously in Canada.

Printed in the United States of America.

International Standard Book Number: 0-669-19540-5

4 5 6 7 8 9 0

TO THE STUDENT

The *Study and Solutions Guide for Trigonometry* is a supplement to the text by Roland E. Larson and Robert P. Hostetler.

As a mathematics instructor, I often have students come to me with questions about the assigned homework. When I ask to see their work, the reply often is "I didn't know where to start." The purpose of the *Study Guide* is to provide brief summaries of the topics covered in the textbook and enough detailed solutions to problems so that you will be able to work the remaining exercises.

A special thanks to Linda M. Bollinger for typing this guide. Also I would like to thank my husband Edward L. Schlindwein for his support during the several months I worked on this project.

If you have any corrections or suggestions for improving this *Study Guide*, I would appreciate hearing from you.

Good luck with your study of trigonometry.

Dianna L. Zook
The Pennsylvania State University
Erie, Pennsylvania 16563

CONTENTS

CHAPTER 1

Prerequisites for Trigonometry

SECTION 1.1

The Real Number System

- You should be able to plot real numbers on the real number line.

- You should know inequality notation. ($<$, $\leq$, $>$, $\geq$)

- You should know the definition of absolute value.

$$|a| = \begin{cases} a, & \text{if } a \geq 0 \\ -a, & \text{if } a < 0 \end{cases}$$

- You should know the following properties of absolute value.
  (a) $|a| \geq 0$
  (b) $|-a| = |a|$
  (c) $|ab| = |a||b|$
  (d) $\left|\dfrac{a}{b}\right| = \dfrac{|a|}{|b|}, \quad b \neq 0$

- The distance between the two points a and b on the real number line is:

$$d(a, b) = |b - a| = |a - b|.$$

- There are two basic types of absolute value inequalities.
  (a) $|x| < a \Rightarrow -a < x < a$
  (b) $|x| > a \Rightarrow x < -a \quad \text{OR} \quad x > a$

Solutions to Selected Exercises

5. Plot the two real numbers $\frac{5}{6}$ and $\frac{2}{3}$ on the number real line and place the appropriate inequality sign between them.

Solution:

$$\frac{5}{6} > \frac{2}{3}$$

7. Use inequality notation to denote the expression "x is negative".

Solution:
"x is negative" can be written as $x < 0$.

15. Write $|3 - \pi|$ without absolute value signs.

Solution:
Since $3 < \pi$, $3 - \pi$ is negative. Therefore, $|3 - \pi| = -(3 - \pi) = \pi - 3$.

19. Write $-3 - |-3|$ without absolute value signs.

Solution:

$$-3 - |-3| = -3 - [-(-3)] = -3 - 3 = -6$$

23. Find the distance between the points $-\frac{5}{2}$ and 0 on the real line.

Solution:

$$d\left(-\frac{5}{2}, \ 0\right) = \left|0 - \left(-\frac{5}{2}\right)\right| = \left|\frac{5}{2}\right| = \frac{5}{2}$$

27. Find the distance between the points 9.34 and -5.65 on the real line.

Solution:

$$\begin{aligned}
d(9.34, \ -5.65) &= |9.34 - (-5.65)| \\
&= |9.34 + 5.65| \\
&= |14.99| = 14.99
\end{aligned}$$

31. Use absolute value notation to describe the expression "the distance between z and $\frac{3}{2}$ is greater than 1."

Solution:
Since

$$d(z, \tfrac{3}{2}) = |z - \tfrac{3}{2}| \quad \text{and} \quad d(z, \tfrac{3}{2}) > 1$$

we have

$$|z - \tfrac{3}{2}| > 1.$$

37. (a) Use a calculator to order the following real numbers, from smallest to largest.

$$\frac{7071}{5000}, \ \frac{584}{413}, \ \sqrt{2}, \ \frac{47}{33}, \ \frac{127}{90}$$

(b) Which of the rational numbers in part (a) is closest to $\sqrt{2}$?

Solution:

(a) $\dfrac{7071}{5000} = 1.4142$

$\dfrac{584}{413} = 1.414043584$

$\sqrt{2} = 1.414213562$

$\dfrac{47}{33} = 1.42\overline{42}$

$\dfrac{127}{90} = 1.41\overline{1}$

$\dfrac{127}{90} < \dfrac{584}{413} < \dfrac{7071}{5000} < \sqrt{2} < \dfrac{47}{33}$

(b) $\dfrac{7071}{5000}$ is closest to $\sqrt{2}$.

43. Match $|x| < 4$ with its graph.

Solution:

$|x| < 4$

$-4 < x < 4$ \qquad Matches graph (g)

49. Use absolute value notation to define the pair of intervals on the real line.

Solution:

$|x - 9| \geq 3$

53. Use absolute value notation to define following the interval on the real line: All real numbers whose distances from -3 are more than 5.

Solution:

$|x - (-3)| > 5$

$|x + 3| > 5$

SECTION 1.2

Solving Equations

- To solve an equation you may:

  (a) Remove symbols of grouping, combine like terms, or reduce fractions on both sides of the equation.

  (b) Add or subtract the same quantity to both sides of the equation.

  (c) Multiply or divide both sides of the equation by the same nonzero quantity.

  (d) Interchange the two sides of the equation.

- Be sure to check your answers for extraneous solutions.

- You should be able to factor special polynomial forms.

  (a) $u^2 - v^2 = (u + v)(u - v)$

  (b) $u^2 + 2uv + v^2 = (u + v)^2$

  (c) $u^2 - 2uv + v^2 = (u - v)^2$

  (d) $u^3 + v^3 = (u + v)(u^2 - uv + v^2)$

  (e) $u^3 - v^3 = (u - v)(u^2 + uv + v^2)$

- You should know the Quadratic Formula. For $ax^2 + bx + c = 0$, $a \neq 0$, the solutions are

$$x = \frac{-b \pm \sqrt{b^2 - 4ac}}{2a}.$$

- Use the discriminant to determine the type of solutions of a quadratic equation.

  (a) If $b^2 - 4ac > 0$, there are two distinct real solutions.

  (b) If $b^2 - 4ac = 0$, there is one repeated real solution.

  (c) If $b^2 - 4ac < 0$, there are no real solutions.

- Be able to solve polynomial equations of higher degree.

  (a) Factor into linear and quadratic factors.

  (b) Use the Quadratic Formula on polynomials that are similar to quadratics.

  (c) Factor by grouping.

Solutions to Selected Exercises

3. Determine whether the given value of x is a solution of the equation $3x^2 + 2x - 5 = 2x^2 - 2$.

    (a) $x = -3$ (b) $x = 1$

    (c) $x = 4$ (d) $x = -5$

Solution:

(a) $3(-3)^2 + 2(-3) - 5 \stackrel{?}{=} 2(-3)^2 - 2$ (b) $3(1)^2 + 2(1) - 5 \stackrel{?}{=} 2(1)^2 - 2$

$$16 = 16$$

$$0 = 0$$

        $x = -3$ is a solution. $x = 1$ is a solution.

(c) $3(4)^2 + 2(4) - 5 \stackrel{?}{=} 2(4)^2 - 2$ (d) $3(-5)^2 + 2(-5) - 5 \stackrel{?}{=} 2(-5)^2 - 2$

$$51 \neq 30$$

$$60 \neq 48$$

        $x = 4$ is not a solution. $x = -5$ is not a solution.

5. Solve the equation $2(x + 5) - 7 = 3(x - 2)$.

Solution:

$$2(x + 5) - 7 = 3(x - 2)$$
$$2x + 10 - 7 = 3x - 6$$
$$2x + 3 = 3x - 6$$
$$-x + 3 = -6$$
$$-x = -9$$
$$x = 9$$

7. Solve the equation $\dfrac{5x}{4} + \dfrac{1}{2} = x - \dfrac{1}{2}$.

Solution:

$$\frac{5x}{4} + \frac{1}{2} = x - \frac{1}{2}$$
$$4\left(\frac{5x}{4} + \frac{1}{2}\right) = 4\left(x - \frac{1}{2}\right)$$
$$5x + 2 = 4x - 2$$
$$x + 2 = -2$$
$$x = -4$$

9. Solve the equation $0.25x + 0.75(10 - x) = 3$.

Solution:

$$0.25x + 0.75(10 - x) = 3$$
$$100[0.25x + 0.75(10 - x)] = 100(3)$$
$$25x + 75(10 - x) = 300$$
$$25x + 750 - 75x = 300$$
$$-50x + 750 = 300$$
$$-50x = -450$$
$$x = 9$$

11. Solve the equation $x + 8 = 2(x - 2) - x$, if possible.

Solution:

$$x + 8 = 2(x - 2) - x$$
$$x + 8 = 2x - 4 - x$$
$$x + 8 = x - 4$$
$$8 = -4 \qquad \text{Not possible}$$

Thus, the equation has no solution.

15. Solve the equation

$$\frac{5x - 4}{5x + 4} = \frac{2}{3}.$$

Solution:

$$\frac{5x - 4}{5x + 4} = \frac{2}{3}$$
$$3(5x - 4) = 2(5x + 4) \qquad \text{Cross multiply}$$
$$15x - 12 = 10x + 8$$
$$5x = 20$$
$$x = 4$$

19. Solve the equation

$$\frac{1}{x - 3} + \frac{1}{x + 3} = \frac{10}{x^2 - 9}.$$

Solution:

$$\frac{1}{x-3} + \frac{1}{x+3} = \frac{10}{x^2-9}$$

$$\frac{(x+3)+(x-3)}{(x-3)(x+3)} = \frac{10}{x^2-9}$$

$$(x^2-9)\left(\frac{2x}{x^2-9}\right) = \left(\frac{10}{x^2-9}\right)(x^2-9)$$

$$2x = 10$$

$$x = 5$$

21. Solve the equation

$$\frac{7}{2x+1} - \frac{8x}{2x-1} = -4 \, .$$

Solution:

$$\frac{7}{2x+1} - \frac{8x}{2x-1} = -4$$

$$(2x+1)(2x-1)\left[\frac{7}{2x+1} - \frac{8x}{2x-1}\right] = -4(2x+1)(2x-1)$$

$$7(2x-1) - 8x(2x+1) = -4(4x^2-1)$$

$$14x - 7 - 16x^2 - 8x = -16x^2 + 4$$

$$-16x^2 + 6x - 7 = -16x^2 + 4$$

$$6x - 7 = 4$$

$$6x = 11$$

$$x = \frac{11}{6}$$

23. Solve the equation $(x+2)^2 + 5 = (x+3)^2$.

Solution:

$$(x+2)^2 + 5 = (x+3)^2$$

$$x^2 + 4x + 4 + 5 = x^2 + 6x + 9$$

$$4x + 9 = 6x + 9$$

$$4x = 6x$$

$$-2x = 0$$

$$x = 0$$

29. Solve $x^2 - 2x - 8 = 0$ by factoring.

Solution:

$$x^2 - 2x - 8 = 0$$
$$(x + 2)(x - 4) = 0$$
$$x = -2 \quad \text{or} \quad x = 4$$

33. Solve $2x^2 = 19x + 33$ by factoring.

Solution:

$$2x^2 = 19x + 33$$
$$2x^2 - 19x - 33 = 0$$
$$(2x + 3)(x - 11) = 0$$
$$2x + 3 = 0 \quad \text{or} \quad x - 11 = 0$$
$$2x = -3 \qquad\qquad x = 11$$
$$x = -\tfrac{3}{2}$$

35. Solve $3x^2 = 36$ by taking the square root of both sides.

Solution:

$$3x^2 = 36$$
$$x^2 = 12$$
$$x = \pm\sqrt{12}$$
$$x = \pm 2\sqrt{3}$$

41. Use the Quadratic Formula to solve $16x^2 + 8x - 3 = 0$.

Solution:
$16x^2 + 8x - 3 = 0; \quad a = 16, \ b = 8, \ c = -3$

$$x = \frac{-8 \pm \sqrt{8^2 - 4(16)(-3)}}{2(16)} = \frac{-8 \pm \sqrt{256}}{32} = \frac{-8 \pm 16}{32}$$

$$x = \frac{-8 + 16}{32} = \frac{1}{4}$$

$$x = \frac{-8 - 16}{32} = -\frac{3}{4}$$

47. Use the Quadratic Formula to solve $4x^2 + 4x = 7$.

Solution:

$$4x^2 + 4x = 7$$
$$4x^2 + 4x - 7 = 0$$
$$a = 4, \ b = 4, \ c = -7$$

$$x = \frac{-4 \pm \sqrt{(4)^2 - 4(4)(-7)}}{2(4)} = \frac{-4 \pm \sqrt{16 + 112}}{8}$$

$$= \frac{-4 \pm \sqrt{128}}{8} = \frac{-4 \pm 8\sqrt{2}}{8} = \frac{4(-1 \pm 2\sqrt{2})}{8}$$

$$= \frac{-1 \pm 2\sqrt{2}}{2} = -\frac{1}{2} \pm \sqrt{2}$$

51. Use the Quadratic Formula to solve

$$\frac{1}{x} - \frac{1}{x+1} = 3.$$

Solution:

$$\frac{1}{x} - \frac{1}{x+1} = 3$$

$$\frac{(x+1) - x}{x(x+1)} = 3$$

$$\frac{1}{x^2 + x} = 3$$

$$1 = 3(x^2 + x)$$

$$0 = 3x^2 + 3x - 1; \ a = 3, \ b = 3, \ c = -1$$

$$x = \frac{-3 \pm \sqrt{(3)^2 - 4(3)(-1)}}{2(3)} = \frac{-3 \pm \sqrt{21}}{6} = -\frac{1}{2} \pm \frac{\sqrt{21}}{6}$$

53. Use a calculator to solve $5.1x^2 - 1.7x - 3.2 = 0$. Round your answer to three decimal places.

Solution:

$$5.1x^2 - 1.7x - 3.2 = 0; \ a = 5.1, \ b = -1.7, \ c = -3.2$$

$$x = \frac{-(-1.7) \pm \sqrt{(-1.7)^2 - 4(5.1)(-3.2)}}{2(5.1)}$$

$$x = \frac{1.7 \pm \sqrt{68.17}}{10.2} \approx \frac{1.7 \pm 8.2565}{10.2}$$

$$x \approx 0.976 \quad \text{or} \quad x \approx -0.643$$

61. Find all solutions of $x^4 - 81 = 0$.

Solution:

$$x^4 - 81 = 0$$
$$(x^2 + 9)(x^2 - 9) = 0$$
$$(x^2 + 9)(x + 3)(x - 3) = 0$$
$$x = \pm 3$$

65. Find all solutions of $x^3 - 3x^2 - x + 3 = 0$.

Solution:

$$x^3 - 3x^2 - x + 3 = 0$$
$$x^2(x - 3) - (x - 3) = 0$$
$$(x - 3)(x^2 - 1) = 0$$
$$(x - 3)(x + 1)(x - 1) = 0$$
$$x = 3 \quad \text{or} \quad x = \pm 1$$

71. Find all solutions of $x^4 + 5x^2 - 36 = 0$.

Solution:

$$x^4 + 5x^2 - 36 = 0$$
$$(x^2 + 9)(x^2 - 4) = 0$$
$$(x^2 + 9)(x + 2)(x - 2) = 0$$
$$x = \pm 2$$

75. Find all solutions of $x^6 + 7x^3 - 8 = 0$.

Solution:

$$x^6 + 7x^3 - 8 = 0$$
$$(x^3 + 8)(x^3 - 1) = 0$$
$$(x + 2)(x^2 - 2x + 4)(x - 1)(x^2 + x + 1) = 0$$
$$x = -2, \quad x = \frac{2 \pm \sqrt{4 - 16}}{2}, \quad x = 1 \quad \text{or} \quad x = \frac{-1 \pm \sqrt{1 - 4}}{2}$$
$$x = -2, \qquad\qquad\qquad\qquad x = 1$$

SECTION 1.3

The Cartesian Plane

■ You should be able to plot points.

■ You should know that the distance between $(x_1,\ y_1)$ and $(x_2,\ y_2)$ in the plane is

$$d = \sqrt{(x_2 - x_1)^2 + (y_2 - y_1)^2}.$$

■ You should know that the midpoint of the line segment joining $(x_1,\ y_1)$ and $(x_2,\ y_2)$ is

$$\left(\frac{x_1 + x_2}{2},\ \frac{y_1 + y_2}{2}\right).$$

Solutions to Selected Exercises

3. Sketch the square with vertices $(2,\ 4)$, $(5,\ 1)$, $(2,\ -2)$, and $(-1,\ 1)$.

Solution:

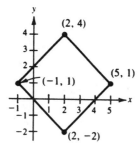

7. Find the distance between the points $(-3,\ -1)$ and $(2,\ -1)$.

Solution:

Since the points $(-3,\ -1)$ and $(2,\ -1)$ lie on a vertical line, the distance between the points is given by the absolute value of the difference of their x–coordinates.

$$d = |-3 - 2| = 5$$

11. For the indicated triangle (a) find the length of the two sides of the right triangle and use the Pythagorean Theorem to find the length of the hypotenuse, and (b) use the Distance Formula to find the length of the hypotenuse of the triangle.

Solution:

(a) $a = |-3 - 7| = 10$

$b = |4 - 1| = 3$

$c = \sqrt{10^2 + 3^2} = \sqrt{109}$

(b) $c = \sqrt{(7 - (-3))^2 + (4 - 1)^2}$

$= \sqrt{10^2 + 3^2}$

$= \sqrt{109}$

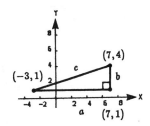

15. (a) Plot the points $(-4, 10)$ and $(4, -5)$, (b) find the distance between the points, and (c) find the midpoint of the line segment joining the points.

Solution:

(a)

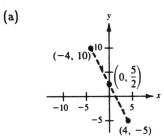

(b) $d = \sqrt{(-4 - 4)^2 + (10 - (-5))^2}$

$= \sqrt{(-8)^2 + (15)^2}$

$= \sqrt{289}$

$= 17$

(c) $m = \left(\dfrac{-4 + 4}{2}, \dfrac{10 + (-5)}{2}\right)$

$= \left(0, \dfrac{5}{2}\right)$

21. (a) Plot the points $(6.2, 5.4)$, and $(-3.7, 1.8)$, (b) find the distance between the points, and (c) find the midpoint of the line segment joining the points.

Solution:

(a)

(b) $d = \sqrt{(6.2 - (-3.7))^2 + (5.4 - 1.8)^2}$

$= \sqrt{(9.9)^2 + (3.6)^2}$

$= \sqrt{110.97}$

≈ 10.5342

(c) $m = \left(\dfrac{6.2 + (-3.7)}{2}, \dfrac{5.4 + 1.8}{2}\right)$

$= (1.25, \ 3.6)$

25. Show that the points $(4, 0)$, $(2, 1)$, and $(-1, -5)$ form the vertices of a right triangle.

Solution:

$$d_1 = \sqrt{(-1-2)^2 + (-5-1)^2} = \sqrt{45}$$
$$d_2 = \sqrt{(2-4)^2 + (1-0)^2} = \sqrt{5}$$
$$d_3 = \sqrt{(4-(-1))^2 + (0-(-5))^2} = \sqrt{50}$$

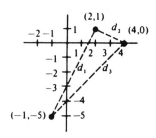

Since $d_1{}^2 + d_2{}^2 = d_3{}^2$, we can conclude by the Pythagorean Theorem that the triangle is a right triangle.

29. Find x so that the distance between $(1, 2)$ and $(x, -10)$ is 13.

Solution:

$$\sqrt{(x-1)^2 + (-10-2)^2} = 13$$
$$\sqrt{x^2 - 2x + 1 + 144} = 13$$
$$x^2 - 2x + 145 = 169$$
$$x^2 - 2x - 24 = 0$$
$$(x+4)(x-6) = 0$$
$$x = -4 \quad \text{or} \quad x = 6$$

33. Find a relationship between x and y so that (x, y) is equidistant from the points $(4, -1)$ and $(-2, 3)$.

Solution:

The distance between $(4, -1)$ and (x, y) is equal to the distance between $(-2, 3)$ and (x, y).

$$\sqrt{(x-4)^2 + (y+1)^2} = \sqrt{(x+2)^2 + (y-3)^2}$$
$$(x-4)^2 + (y+1)^2 = (x+2)^2 + (y-3)^2$$
$$x^2 - 8x + 16 + y^2 + 2y + 1 = x^2 + 4x + 4 + y^2 - 6y + 9$$
$$4 = 12x - 8y$$
$$1 = 3x - 2y$$
$$2y = 3x - 1$$

37. Determine the quadrant in which (x, y) is located so that the conditions $x > 0$ and $y > 0$ are satisfied.

Solution:

$x > 0 \rightarrow x$ lies in Quadrant I or in Quadrant IV

$y > 0 \rightarrow y$ lies in Quadrant I or in Quadrant II

$x > 0$ and $y > 0 \rightarrow (x, y)$ lies in Quadrant I

45. Use the Midpoint Formula twice to find the three points that divide the line segment joining (x_1, y_1) and (x_2, y_2) into four parts.

Solution:

The midpoint of the given line segment is

$$\left(\frac{x_1 + x_2}{2}, \frac{y_1 + y_2}{2}\right).$$

The midpoint between (x_1, y_1) and $\left(\dfrac{x_1 + x_2}{2}, \dfrac{y_1 + y_2}{2}\right)$ is

$$\left(\frac{x_1 + \dfrac{x_1 + x_2}{2}}{2}, \frac{y_1 + \dfrac{y_1 + y_2}{2}}{2}\right) = \left(\frac{3x_1 + x_2}{4}, \frac{3y_1 + y_2}{4}\right).$$

The midpoint between $\left(\dfrac{x_1 + x_2}{2}, \dfrac{y_1 + y_2}{2}\right)$ and (x_2, y_2) is

$$\left(\frac{\dfrac{x_1 + x_2}{2} + x_2}{2}, \frac{\dfrac{y_1 + y_2}{2} + y_2}{2}\right) = \left(\frac{x_1 + 3x_2}{4}, \frac{y_1 + 3y_2}{4}\right)$$

Thus, the three points are

$$\left(\frac{3x_1 + x_2}{4}, \frac{3y_1 + y_2}{4}\right), \quad \left(\frac{x_1 + x_2}{2}, \frac{y_1 + y_2}{2}\right), \quad \text{and} \quad \left(\frac{x_1 + 3x_2}{4}, \frac{y_1 + 3y_2}{4}\right).$$

47. Use the Midpoint Formula to estimate the sales of a company for 1983, given the sales in 1980 and 1986. Assume the annual sales followed a linear pattern.

Year	1980	1986
Sales	$520,000	$740,000

Solution:

$$\frac{520,000 + 740,000}{2} = 630,000$$

The estimated sales for 1983 is $630,000.

SECTION 1.4

Graphs of Equations

■ You should be able to use the point-plotting method of graphing.

(a) Isolate one of the variables, if possible.
(b) Make a table of solution points.
(c) Plot these points.
(d) Connect the points with a smooth curve.

■ Be able to find the intercepts.

(a) To find the x-intercept(s), let $y = 0$ and solve for x.
(b) To find the y-intercept(s), let $x = 0$ and solve for y.

■ Be able to determine if the graph has any symmetries.

(a) To test for y-axis symmetry, replace x with $-x$.
(b) To test for x-axis symmetry, replace y with $-y$.
(c) To test for origin symmetry, replace x with $-x$ *and* y with $-y$.

If the substitution yields an equivalent equation, then the graph has that type of symmetry.

■ The standard equation of a circle with center $(h, \; k)$ and radius r is:

$$(x - h)^2 + (y - k)^2 = r^2.$$

■ The general equation of a circle is:

$$Ax^2 + Ay^2 + Dx + Ey + F = 0, \; A \neq 0.$$

Solutions to Selected Exercises

5. Determine whether the points (a) $\left(1, \; \frac{1}{5}\right)$, and (b) $\left(2, \; \frac{1}{2}\right)$ lie on the graph of the equation $x^2 y - x^2 + 4y = 0$.

Solution:

(a) $\left(1, \; \frac{1}{5}\right)$ lies on the graph since $(1)^2 \left(\frac{1}{5}\right) - (1)^2 + 4\left(\frac{1}{5}\right) = \frac{1}{5} - 1 + \frac{4}{5} = 0$.

(b) $\left(2, \; \frac{1}{2}\right)$ lies on the graph since $(2)^2 \left(\frac{1}{2}\right) - (2)^2 + 4\left(\frac{1}{2}\right) = 2 - 4 + 2 = 0$.

7. Find the constant C so that the ordered pair $(2, 6)$ is a solution point of the equation $y = x^2 + C$.

Solution:

$$y = x^2 + C$$
$$6 = (2)^2 + C$$
$$6 = 4 + C$$
$$C = 2$$

13. Find the x- and y-intercepts of the graph of the equation $y = x^2 + x - 2$.

Solution:

Let $y = 0$. Then $0 = x^2 + x - 2 = (x + 2)(x - 1)$ and $x = -2$ or $x = 1$.

x-intercepts: $(-2, 0)$ and $(1, 0)$

Let $x = 0$. Then $y = -2$.

y-intercept: $(0, -2)$

17. Find the x- and y-intercepts of the graph of the equation $xy - 2y - x + 1 = 0$.

Solution:

Let $y = 0$. Then $-x + 1 = 0$ and $x = 1$.

x-intercept: $(1, 0)$

Let $x = 0$. Then $-2y + 1 = 0$ and $y = \frac{1}{2}$.

y-intercept: $\left(0, \frac{1}{2}\right)$

21. Check for symmetry with respect to both axes and the origin for $x - y^2 = 0$.

Solution:

By replacing y with $-y$, we have

$$x - (-y)^2 = 0$$
$$x - y^2 = 0$$

which is the original equation. Replacing x with $-x$ or replacing both x and y with $-x$ and $-y$ does not yield equivalent equations. Thus, $x - y^2 = 0$ is symmetric with respect to the x-axis.

25. Check for symmetry with respect to both axes and the origin for

$$y = \frac{x}{x^2 + 1}.$$

Solution:

Replacing x with $-x$ or y with $-y$ does not yield equivalent equations. Replacing x with $-x$ and y with $-y$ yields

$$-y = \frac{-x}{(-x)^2 + 1}$$

$$-y = \frac{-x}{x^2 + 1} \qquad \text{Multiply both sides by } -1$$

$$y = \frac{x}{x^2 + 1}$$

Thus, $y = \dfrac{x}{x^2 + 1}$ is symmetric with respect to the origin.

Note: An equation is symmetric with respect to the origin if it is symmetric with respect to both the x-axis and the y-axis. Also, if an equation is symmetric with respect to the origin, then one of the following is true:

1. The equation has both x-axis and y-axis symmetry or
2. The equation has neither x-axis nor y-axis symmetry.

31. Match $y = x^3 - x$ with its graph.

Solution:

$y = x^3 - x$
x-intercepts: $(-1, \ 0)$, $(0, \ 0)$, $(1, \ 0)$
y-intercept: $(0, \ 0)$
Symmetry: Origin
Matches graph (e)

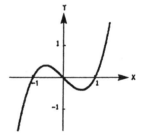

33. Sketch the graph of $y = -3x + 2$. Identify any intercepts and test for symmetry.

Solution:

$y = -3x + 2$
x-intercept: $\left(\frac{2}{3}, \ 0\right)$
y-intercept: $(0, \ 2)$
No symmetries

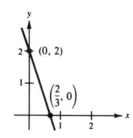

37. Sketch the graph of $y = x^2 - 4x + 3$. Identify any intercepts and test for symmetry.

Solution:

$y = x^2 - 4x + 3 = (x - 1)(x - 3)$
x-intercepts: $(1, 0)$, $(3, 0)$
y-intercept: $(0, 3)$
No symmetries

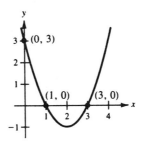

43. Sketch the graph of $y = \sqrt{x - 3}$. Identify any intercepts and test for symmetry.

Solution:

$y = \sqrt{x - 3}$
x-intercept: $(3, 0)$
No y-intercept
No symmetry

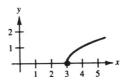

x	3	4	7	12
y	0	1	2	3

Note: The domain is $[3, \infty)$ and the range is $[0, \infty)$.

49. Sketch the graph of $x = y^2 - 1$. Identify any intercepts and test for symmetry.

Solution:

$x = y^2 - 1$
x-intercept: $(-1, 0)$
y-intercepts: $(0, -1)$, $(0, 1)$
x-axis symmetry

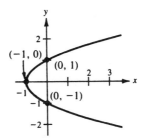

55. Find the standard form of the equation of the circle with center $(2, -1)$ and radius 4.

Solution:

$$(x - 2)^2 + (y + 1)^2 = 4^2$$
$$x^2 - 4x + 4 + y^2 + 2y + 1 = 16$$
$$x^2 + y^2 - 4x + 2y - 11 = 0$$

57. Find the standard form of the equation of the circle with center $(-1, \ 2)$ and solution point $(0, \ 0)$.

Solution:

$$(x+1)^2 + (y-2)^2 = r^2$$
$$(0+1)^2 + (0-2)^2 = r^2 \quad \rightarrow \quad r^2 = 5$$
$$(x+1)^2 + (y-2)^2 = 5$$
$$x^2 + y^2 + 2x - 4y = 0$$

61. Write the following equation of the circle in standard form and sketch its graph.

$$x^2 + y^2 - 2x + 6y + 6 = 0$$

Solution:

$$x^2 + y^2 - 2x + 6y + 6 = 0$$
$$(x^2 - 2x + 1) + (y^2 + 6y + 9) = -6 + 1 + 9$$
$$(x-1)^2 + (y+3)^2 = 4$$

Center: $(1, \ -3)$
Radius: 2

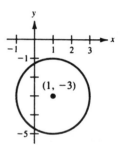

63. Write the following equation of the circle in standard form and sketch its graph.

$$x^2 + y^2 - 2x + 6y + 10 = 0$$

Solution:

$$x^2 + y^2 - 2x + 6y + 10 = 0$$
$$(x^2 - 2x + 1) + (y^2 + 6y + 9) = -10 + 1 + 9$$
$$(x-1)^2 + (y+3)^2 = 0$$

Graph is the point $(1, \ -3)$.

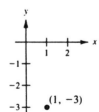

67. Write the following equation of the circle in standard form and sketch its graph.

$$16x^2 + 16y^2 + 16x + 40y - 7 = 0$$

Solution:

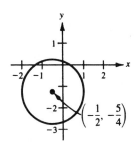

$$16x^2 + 16y^2 + 16x + 40y - 7 = 0$$

$$16\left(x^2 + x + \tfrac{1}{4}\right) + 16\left(y^2 + \tfrac{5}{2}y + \tfrac{25}{16}\right) = 7 + 4 + 25$$

$$16\left(x + \tfrac{1}{2}\right)^2 + 16\left(y + \tfrac{5}{4}\right)^2 = 36$$

$$\left(x + \tfrac{1}{2}\right)^2 + \left(y + \tfrac{5}{4}\right)^2 = \tfrac{9}{4}$$

Center: $\left(-\tfrac{1}{2}, -\tfrac{5}{4}\right)$

Radius: $\tfrac{3}{2}$

73. Use the Distance Formula to show that the points shown in the figure divide the unit circle into four equal parts.

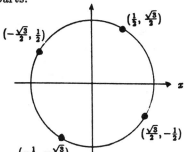

Solution:

Let $A = \left(\dfrac{1}{2}, \dfrac{\sqrt{3}}{2}\right)$, $B = \left(-\dfrac{\sqrt{3}}{2}, \dfrac{1}{2}\right)$, $C = \left(-\dfrac{1}{2}, -\dfrac{\sqrt{3}}{2}\right)$, and $D = \left(\dfrac{\sqrt{3}}{2}, -\dfrac{1}{2}\right)$.

Then, $d(AB) = \sqrt{\left(-\dfrac{\sqrt{3}}{2} - \dfrac{1}{2}\right)^2 + \left(\dfrac{1}{2} - \dfrac{\sqrt{3}}{2}\right)^2}$

$$= \sqrt{\left(\dfrac{-\sqrt{3}-1}{2}\right)^2 + \left(\dfrac{1-\sqrt{3}}{2}\right)^2}$$

$$= \sqrt{\dfrac{3 + 2\sqrt{3} + 1}{4} + \dfrac{1 - 2\sqrt{3} + 3}{4}} = \sqrt{\dfrac{8}{4}} = \sqrt{2}$$

In a similar manner, we can show that $d(BC) = d(CD) = d(DA) = \sqrt{2}$ as well. Therefore, these points divide the unit circle into four equal parts.

SECTION 1.5

Functions

■ Given an equation, you should be able to determine if it represents a function.

■ Given a function, you should be able to do the following.
  (a) Find the domain.
  (b) Find the range.
  (c) Determine if it is one-to-one.
  (d) Evaluate it at specific values.

Solutions to Selected Exercises

5. Evaluate the function at the specified value of the independent variable and simplify the results.

$$f(x) = 2x - 3$$

(a) $f(1)$ (b) $f(-3)$
(c) $f(x-1)$ (d) $f(1/4)$

Solution:

(a) $f(1) = 2(1) - 3 = -1$ (b) $f(-3) = 2(-3) - 3 = -9$
(c) $f(x-1) = 2(x-1) - 3 = 2x - 5$ (d) $f(1/4) = 2(1/4) - 3 = -5/2$

9. Evaluate the function at the specified value of the independent variable and simplify the results.

$$f(y) = 3 - \sqrt{y}$$

(a) $f(4)$ (b) $f(100)$
(c) $f(4x^2)$ (d) $f(0.25)$

Solution:

(a) $f(4) = 3 - \sqrt{4} = 1$ (b) $f(100) = 3 - \sqrt{100} = -7$
(c) $f(4x^2) = 3 - \sqrt{4x^2} = 3 - 2|x|$ (d) $f(0.25) = 3 - \sqrt{0.25} = 2.5$

13. Evaluate the function at the specified value of the independent variable and simplify the results.

$$f(x) = \frac{|x|}{x}$$

(a) $f(2)$ (b) $f(-2)$

(c) $f(x^2)$ (d) $f(x-1)$

Solution:

(a) $f(2) = \dfrac{|2|}{2} = 1$ (b) $f(-2) = \dfrac{|-2|}{-2} = -1$

(c) $f(x^2) = \dfrac{|x^2|}{x^2} = 1$ (d) $f(x-1) = \dfrac{|x-1|}{x-1}$

15. Evaluate the function at the specified value of the independent variable and simplify the results.

$$f(x) = \begin{cases} 2x+1, & x < 0 \\ 2x+2, & x \geq 0 \end{cases}$$

(a) $f(-1)$ (b) $f(0)$

(c) $f(1)$ (d) $f(2)$

Solution:

(a) $f(-1) = 2(-1) + 1 = -1$ (b) $f(0) = 2(0) + 2 = 2$

(c) $f(1) = 2(1) + 2 = 4$ (d) $f(2) = 2(2) + 2 = 6$

17. For $f(x) = x^2 - x + 1$, find

$$\frac{f(2+h) - f(2)}{h}$$

and simplify your answer.

Solution:

$$f(x) = x^2 - x + 1$$
$$f(2+h) = (2+h)^2 - (2+h) + 1$$
$$= 4 + 4h + h^2 - 2 - h + 1$$
$$= h^2 + 3h + 3$$
$$f(2) = (2)^2 - 2 + 1 = 3$$
$$f(2+h) - f(2) = h^2 + 3h$$
$$\frac{f(2+h) - f(2)}{h} = h + 3$$

25. Find all real values x such that $f(x) = 0$ for $f(x) = x^2 - 9$.

Solution:

$$x^2 - 9 = 0$$
$$x^2 = 9$$
$$x = \pm 3$$

27. Find all real values x such that $f(x) = 0$ for

$$f(x) = \frac{3}{x-1} + \frac{4}{x-2}.$$

Solution:

$$\frac{3}{x-1} + \frac{4}{x-2} = 0$$
$$3(x-2) + 4(x-1) = 0$$
$$7x - 10 = 0$$
$$x = \frac{10}{7}$$

31. Find the domain of $h(t) = 4/t$.

Solution:

The domain includes all real numbers except 0, i.e. $t \neq 0$.

35. Find the domain of $f(x) = \sqrt[4]{1 - x^2}$.

Solution:

Choose x-values for which $1 - x^2 \geq 0$. Using methods of Section 2.8, we find that the domain is $-1 \leq x \leq 1$.

39. Determine if y is a function of x for $x^2 + y^2 = 4$.

Solution:

y is not a function of x since some values of x give two values for y. For example, if $x = 0$, then $y = \pm 2$.

43. Determine if y is a function of x for $2x + 3y = 4$.

Solution:

$$2x + 3y = 4$$
$$y = \tfrac{1}{3}(4 - 2x)$$

y is a function of x.

47. Determine if y is a function of x for $x^2y - x^2 + 4y = 0$.

Solution:

$$x^2y - x^2 + 4y = 0$$
$$y(x^2 + 4) = x^2$$
$$y = \frac{x^2}{x^2 + 4}$$

y is a function of x.

51. Assume that the domain of $f(x) = x^2$ is the set $A = \{-2, -1, 0, 1, 2\}$. Determine the set of ordered pairs representing the function f.

Solution:

$$\{(-2, f(-2)), (-1, f(-1)), (0, f(0)), (1, f(1)), 2, f(2))\}$$
$$\{(-2, 4), (-1, 1), (0, 0), (1, 1), (2, 4)\}$$

57. Find the value(s) of x for which $f(x) = g(x)$ where $f(x) = \sqrt{3x} + 1$ and $g(x) = x + 1$.

Solution:

$$f(x) = g(x)$$
$$\sqrt{3x} + 1 = x + 1$$
$$\sqrt{3x} = x$$
$$3x = x^2$$
$$0 = x^2 - 3x$$
$$0 = x(x - 3)$$
$$x = 0 \quad \text{or} \quad x = 3$$

61. Express the area A of a circle as a function of its circumference C.

Solution:

$$A = \pi r^2, \quad C = 2\pi r$$
$$r = \frac{C}{2\pi}$$
$$A = \pi \left(\frac{C}{2\pi}\right)^2$$
$$A = \frac{C^2}{4\pi}$$

65. A right triangle is formed in the first quadrant by the x- and y-axes and a line through the point $(1, 2)$, as shown in the figure. Write the area of the triangle as a function of x, and determine the domain of the function.

Solution:

$$A = \frac{1}{2}bh = \frac{1}{2}xy$$

Since $(0, \ y)$, $(1, 2)$ and $(x, \ 0)$ all lie on the same line, the slopes between any pair are equal.

$$\frac{2-y}{1-0} = \frac{0-2}{x-1}$$

$$2 - y = -\frac{2}{x-1}$$

$$y = \frac{2}{x-1} + 2$$

$$y = \frac{2x}{x-1}$$

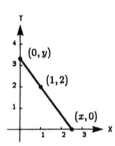

Therefore,

$$A = \frac{1}{2}x\left(\frac{2x}{x-1}\right)$$

$$A = \frac{x^2}{x-1}$$

The domain of A includes x-values such that $x^2/(x-1) > 0$. This results in a domain of $x > 1$.

69. A company produces a product for which the variable cost is \$12.30 per unit and the fixed costs are \$98,000. The product sells for \$17.98. Let x be the number of units produced.

(a) Write the total cost C as a function of the number of units produced.
(b) Write the revenue R as a function of the number of units produced.
(c) Write the profit P as a function of the number of units produced.

    (Note: $P = R - C$.)

Solution:

(a) Cost = variable costs + fixed costs
    $C = 12.30x + 98,000$

(b) Revenue = price per unit × number of units
    $R = 17.98x$

(c) Profit = Revenue − Cost
    $P = 17.98x - (12.30x + 98,000)$
    $P = 5.68x - 98,000$

SECTION 1.6

Graphs of Functions

- ■ You should be able to determine the domain and range of a function from its graph.

- ■ You should be able to use the vertical line test for functions.

- ■ You should know that the graph of $f(x) = c$ is a horizontal line through $(0, c)$.

- ■ You should be able to determine when a function is constant, increasing, or decreasing.

- ■ You should know that f is
  (a) Odd if $f(-x) = -f(x)$.
  (b) Even if $f(-x) = f(x)$.

- ■ You should know the basic types of transformations.

Solutions to Selected Exercises

5. Determine the domain and range of the function $f(x) = \sqrt{25 - x^2}$.

Solution:
From the graph we see that the x-values do
not extend beyond $x = -5$ (on the left) and
$x = 5$ (on the right). The domain is $[-5, 5]$.
Similarly, the y-values do not extend beyond
$y = 0$ and $y = 5$. The range is $[0, 5]$.

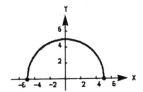

7. Use the vertical line test to determine if y is a function of x where $y = x^2$.

Solution:
Since no vertical line would ever
cross the graph more than one time,
y *is* a function of x.

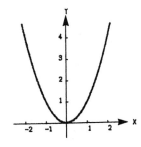

11. Use the vertical line test to determine if y is a function of x where $x^2 = xy - 1$.

Solution:
Since no vertical line would ever
cross the graph more than one time,
y *is* a function of x.

$$y = \frac{x^2 + 1}{x}$$

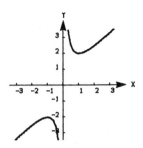

15. (a) Determine the intervals over which the function is increasing, decreasing, or constant, and
(b) determine if the function is even, odd, or neither for $f(x) = x^3 - 3x^2$.

Solution:
(a) By its graph we see that f
is increasing on $(-\infty, 0)$ and
$(2, \infty)$ and is decreasing on $(0, 2)$.

(b) $f(-x) = (-x)^3 - 3(-x)^2$
$$= -x^3 - 3x^2$$
$f(-x) \neq f(x)$ and $f(x) \neq -f(x)$, so
the function is neither odd nor even.

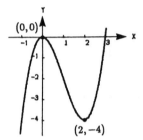

19. (a) Determine the intervals over which the function is increasing, decreasing, or constant, and
(b) determine if the function is even, odd, or neither for $f(x) = x\sqrt{x + 3}$.

Solution:
(a) By its graph we see that f
is increasing on $(-2, \infty)$ and
decreasing on $(-3, -2)$.

(b) $f(-x) = -x\sqrt{-x + 3}$
$f(-x) \neq f(x)$ and $f(-x) \neq -f(x)$, so
the function is neither odd nor even.

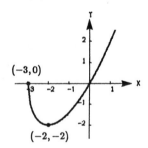

23. Determine whether $g(x) = x^3 - 5x$ is even, odd, or neither.

Solution:

$$g(x) = x^3 - 5x$$
$$g(-x) = (-x)^3 - 5(-x)$$
$$= -x^3 + 5x$$
$$= -(x^3 - 5x)$$
$$= -g(x)$$

Therefore, g is odd.

27. Sketch the graph of $f(x) = 3$ and determine whether the function is odd, even, or neither.

Solution:
$f(x) = 3$
Domain: $(-\infty, \infty)$
Range: $\{3\}$
y-intercept: $(0, 3)$
y-axis symmetry
$f(-x) = 3 = f(x)$
Therefore, f is even.

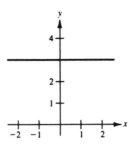

31. Sketch the graph of $g(s) = s^3/4$ and determine whether the function is odd, even, or neither.

Solution:
$g(s) = s^3/4$
Intercept: $(0, 0)$
Origin symmetry
Domain: $(-\infty, \infty)$
Range: $(-\infty, \infty)$
$g(-s) = -g(s)$
Therefore, g is odd.

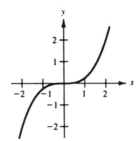

35. Sketch the graph of $g(t) = \sqrt[3]{t} - 1$ and determine whether the function is odd, even, or neither.

Solution:
$g(t) = \sqrt[3]{t} - 1$
x-intercept: $(1, 0)$
y-intercept: $(0, -1)$
Domain: $(-\infty, \infty)$
Range: $(-\infty, \infty)$

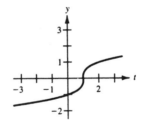

t	0	1	2	-7	9
$g(t)$	-1	0	1	-2	2

$g(-t) = \sqrt[3]{-t} - 1 \neq g(t)$ and $\neq -g(t)$. Therefore, g is neither odd nor even.

37. Sketch the graph of

$$f(x) = \begin{cases} x + 3, & \text{if } x \leq 0 \\ 3, & \text{if } 0 < x \leq 2 \\ 2x - 1, & \text{if } x > 2 \end{cases}$$

and determine whether the function is odd, even, or neither.

Solution:
For $x \leq 0$, $f(x) = x + 3$. For $0 < x \leq 2$, $f(x) = 3$. For $x > 2$, $f(x) = 2x - 1$. Thus, the graph of f is as shown.

$$f(-x) = \begin{cases} -x + 3, & \text{if } x \leq 0 \\ 3, & \text{if } 0 < x \leq 2 \\ -2x - 1, & \text{if } x > 2 \end{cases}$$

So, f is neither odd nor even.

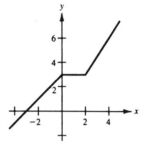

41. Sketch the graph of $f(x) = x^2 - 9$ and determine the interval(s), if any, on the real axis for which $f(x) \geq 0$.

Solution:
$f(x) = x^2 - 9$
x-intercepts: $(-3, 0)$, $(0, 3)$
y-intercept: $(0, -9)$
y-axis symmetry
Domain: $(-\infty, \infty)$
Range: $[-9, \infty)$
$f(x) \geq 0$ on the intervals $(-\infty, -3]$ and $[3, \infty)$.

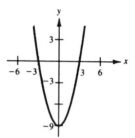

45. Sketch the graph of $f(x) = x^2 + 1$ and determine the interval(s), if any, on the real axis for which $f(x) \geq 0$.

Solution:
$f(x) = x^2 + 1$
x-intercept: None
y-intercept: $(0, 1)$
y-axis symmetry
Domain: $(-\infty, \infty)$
Range: $[1, \infty)$
$f(x) \geq 0$ for all real numbers.

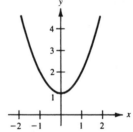

51. Use the graph of $f(x) = \sqrt{x}$ to sketch the graph of each of the following.

(a) $y = \sqrt{x} + 2$
(b) $y = -\sqrt{x}$
(c) $y = \sqrt{x} - 2$
(d) $y = \sqrt{x + 3}$
(e) $y = 2 - \sqrt{x - 4}$
(f) $y = \sqrt{2x}$

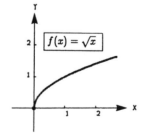

Solution:

(a) $y = \sqrt{x} + 2$
   Vertical shift 2 units upward

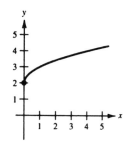

(b) $y = -\sqrt{x}$
   Reflection about the x-axis

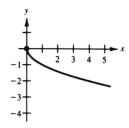

(c) $y = \sqrt{x-2}$
   Horizontal shift 2 units to the right

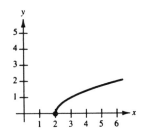

(d) $y = \sqrt{x+3}$
   Horizontal shift 3 units to the left

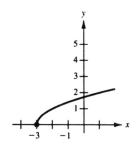

(e) $y = 2 - \sqrt{x-4}$
   Reflection about the x-axis,
   horizontal shift of 4 units to
   the right and a vertical shift
   2 units upward

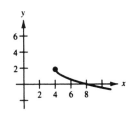

(f) $y = \sqrt{2x}$

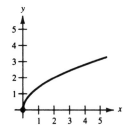

57. Write the height h of the given rectangle as a function of x.

Solution:

$$h = \text{top} - \text{bottom}$$
$$= (4x - x^2) - x^2$$
$$= 4x - 2x^2$$

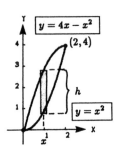

61. Prove that a function of the following form is odd.

$$f(x) = a_{2n+1}x^{2n+1} + a_{2n-1}x^{2n-1} + \ldots + a_3 x^3 + a_1 x$$

Solution:

$$f(x) = a_{2n+1}x^{2n+1} + a_{2n-1}x^{2n-1} + \ldots + a_3 x^3 + a_1 x$$
$$f(-x) = a_{2n+1}(-x)^{2n+1} + a_{2n-1}(-x)^{2n-1} + \ldots + a_3(-x)^3 + a_1(-x)$$
$$= -a_{2n+1}x^{2n+1} - a_{2n-1}x^{2n-1} - \ldots - a_3 x^3 - a_1 x$$
$$= -f(x)$$

Therefore, $f(x)$ is odd.

SECTION 1.7

Combinations of Functions and Inverse Functions

- If $f(x)$ and $g(x)$ are functions, then so are the following.
  - (a) $(f + g)(x) = f(x) + g(x)$
  - (b) $(f - g)(x) = f(x) - g(x)$
  - (c) $(fg)(x) = f(x) \cdot g(x)$
  - (d) $\left(\dfrac{f}{g}\right)(x) = \dfrac{f(x)}{g(x)}, \quad g(x) \neq 0$
  - (e) $(f \circ g)(x) = f(g(x))$
  - (f) $(g \circ f)(x) = g(f(x))$

- f and g are *inverses* of each other if:
  - (a) $f(g(x)) = x$ for every x in the domain of g.
  - (b) $g(f(x)) = x$ for every x in the domain of f.

- f has an inverse if and only if f is one-to-one. That means, if $f(a) = f(b)$, then $a = b$.

- To find f^{-1}, if it exists:
  - (a) Test f to determine if it is one-to-one.
  - (b) Replace $f(x)$ with y.
  - (c) Solve for x.
  - (d) Switch the variables. This is $f^{-1}(x)$.

Solutions to Selected Exercises

3. Given $f(x) = x^2$, $g(x) = 1 - x$, find (a) $(f + g)(x)$, (b) $(f - g)(x)$, (c) $(fg)(x)$, and (d) $(f/g)(x)$. What is the domain of f/g?

Solution:

(a) $(f + g)(x) = f(x) + g(x) = x^2 + (1 - x) = x^2 - x + 1$

(b) $(f - g)(x) = f(x) - g(x) = x^2 - (1 - x) = x^2 + x - 1$

(c) $(fg)(x) = f(x) \cdot g(x) = x^2(1 - x) = x^2 - x^3$

(d) $\left(\dfrac{f}{g}\right)(x) = \dfrac{f(x)}{g(x)} = \dfrac{x^2}{1 - x}$

The domain of f/g is the set of all real numbers except 1, (i.e. $x \neq 1$).

9. Evaluate $(f - g)(2t)$ for $f(x) = x^2 + 1$ and $g(x) = x - 4$.

Solution:

$$
\begin{aligned}
(f - g)(2t) &= f(2t) - g(2t) \\
&= [(2t)^2 + 1] - [(2t) - 4] \\
&= 4t^2 + 1 - 2t + 4 \\
&= 4t^2 - 2t + 5
\end{aligned}
$$

13. Evaluate $(f/g)(5)$ for $f(x) = x^2 + 1$ and $g(x) = x - 4$.

Solution:

$$
\left(\frac{f}{g}\right)(5) = \frac{f(5)}{g(5)} = \frac{(5)^2 + 1}{5 - 4} = 26
$$

19. Given $f(x) = 3x + 5$ and $g(x) = 5 - x$, find (a) $f \circ g$, (b) $g \circ f$, and (c) $f \circ f$.

Solution:

(a) $f \circ g = f\big(g(x)\big)$
$$
\begin{aligned}
&= f(5 - x) \\
&= 3(5 - x) + 5 \\
&= 20 - 3x
\end{aligned}
$$

(b) $g \circ f = g\big(f(x)\big)$
$$
\begin{aligned}
&= g(3x + 5) \\
&= 5 - (3x + 5) \\
&= -3x
\end{aligned}
$$

(c) $f \circ f = f\big(f(x)\big)$
$$
\begin{aligned}
&= f(3x + 5) \\
&= 3(3x + 5) + 5 \\
&= 9x + 20
\end{aligned}
$$

29. (a) Show that $f(x) = x^3$ and $g(x) = \sqrt[3]{x}$ are inverse functions by showing that $f\big(g(x)\big) = x$ and $g\big(f(x)\big) = x$, and (b) graph f and g on the same set of coordinate axes.

Solution:

$$
f(x) = x^3, \quad g(x) = \sqrt[3]{x}
$$

(a) $f(g(x)) = f(\sqrt[3]{x}) = (\sqrt[3]{x})^3 = x$
  $g(f(x)) = g(x^3) = \sqrt[3]{x^3} = x$

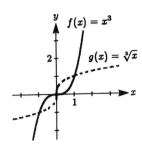

31. (a) Show that $f(x) = \sqrt{x-4}$ and $g(x) = x^2 + 4$, $x \geq 0$ are inverse functions by showing that $f(g(x)) = x$ and $g(f(x)) = x$, and (b) graph f and g on the same set of coordinate axes.

Solution:

$$f(x) = \sqrt{x-4}, \quad g(x) = x^2 + 4, \quad x \geq 0$$

(a) $f(g(x)) = f(x^2 + 4) = \sqrt{(x^2+4)-4} = \sqrt{x^2} = |x| = x, \quad x \geq 0$
  $g(f(x)) = g(\sqrt{x-4}) = (\sqrt{x-4})^2 + 4 = (x-4) + 4 = x$

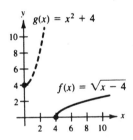

35. Determine whether the function shown is one-to-one.

Solution:
Since the function is decreasing on its entire domain, it is one-to-one.

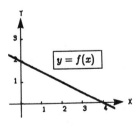

39. Determine whether the function $g(x) = (4 - x)/6$ is one-to-one.

Solution:

Let a and b be real numbers with $g(a) = g(b)$. Then we have

$$\frac{4 - a}{6} = \frac{4 - b}{6}$$

$$4 - a = 4 - b$$

$$-a = -b$$

$$a = b$$

Therefore, $g(x)$ is one-to-one.

43. Determine whether the function $f(x) = -\sqrt{16 - x^2}$ is one-to-one.

Solution:

Since $f(4) = 0$ and $f(-4) = 0$, the function is not one-to-one.

47. Find the inverse of the one-to-one function $f(x) = x^5$. Then graph both f and f^{-1} on the same coordinate plane.

Solution:

$$f(x) = x^5$$

$$y = x^5$$

$$x = \sqrt[5]{y}$$

$$f^{-1}(x) = \sqrt[5]{x}$$

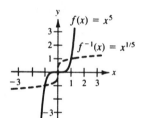

53. Find the inverse of the one-to-one function $f(x) = \sqrt[3]{x - 1}$. Then graph both f and f^{-1} on the same coordinate plane.

Solution:

$$f(x) = \sqrt[3]{x - 1}$$

$$y = \sqrt[3]{x - 1}$$

$$y^3 = x - 1$$

$$x = y^3 + 1$$

$$f^{-1}(x) = x^3 + 1$$

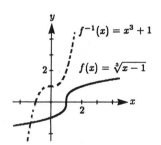

57. Determine whether the function $g(x) = x/8$ is one-to-one. If it is, find its inverse.

Solution:

$$g(a) = g(b)$$
$$\frac{a}{8} = \frac{b}{8}$$
$$a = b \qquad \text{Therefore, } g \text{ is one-to-one.}$$
$$g(x) = \frac{x}{8}$$
$$y = \frac{x}{8}$$
$$x = 8y$$
$$g^{-1}(x) = 8x$$

59. Determine whether the function $p(x) = -4$ is one-to-one. If it is, find its inverse.

Solution:
$p(x) = -4$ for all real numbers x. Therefore, $p(x)$ is not one-to-one.

63. Determine whether the function $h(x) = 1/x$ is one-to-one. If it is, find its inverse.

Solution:

$$h(a) = h(b)$$
$$\frac{1}{a} = \frac{1}{b}$$
$$a = b \qquad \text{Therefore, } h \text{ is one-to-one.}$$
$$h(x) = \frac{1}{x}$$
$$y = \frac{1}{x}$$
$$xy = 1$$
$$x = \frac{1}{y}$$
$$h^{-1}(x) = \frac{1}{x}$$

67. Determine whether the function $g(x) = x^2 - x^4$ is one-to-one. If it is, find its inverse.

Solution:
Since $g(0) = 0$ and $g(1) = 0$, g is not one-to-one.

69. Determine whether the function $f(x) = 25 - x^2$, $x \leq 0$ is one-to-one. If it is, find its inverse.

Solution:

$$f(a) = f(b)$$
$$25 - a^2 = 25 - b^2$$
$$a^2 = b^2 \qquad \text{Since } a,\ b \leq 0, \text{ we have } a = b \text{ and } f \text{ is one-to-one.}$$
$$f(x) = 25 - x^2, \quad x \leq 0$$
$$y = 25 - x^2$$
$$x^2 = 25 - y$$
$$x = -\sqrt{25 - y} \quad \text{Since } x \leq 0$$
$$f^{-1}(x) = -\sqrt{25 - x}$$

73. Use the functions $f(x) = x + 4$ and $g(x) = 2x - 5$ to find $g^{-1} \circ f^{-1}$.

Solution:

$$f(x) = x + 4 \quad \Longrightarrow \quad f^{-1}(x) = x - 4$$
$$g(x) = 2x - 5 \quad \Longrightarrow \quad g^{-1}(x) = \frac{x + 5}{2}$$

$$g^{-1} \circ f^{-1} = g^{-1}(f^{-1}(x))$$
$$= g^{-1}(x - 4)$$
$$= \frac{(x - 4) + 5}{2}$$
$$= \frac{x + 1}{2}$$

REVIEW EXERCISES FOR CHAPTER 1

Solutions to Selected Exercises

5. Write $5 - |-3|$ without absolute value signs.

Solution:

$$5 - |-3| = 5 - [-(-3)] = 5 - 3 = 2$$

11. Graph the inequality, $|x| > 2$, on the real number line.

Solution:
$|x| > 2$ means $x < -2$ or $x > 2$.

15. Use absolute value notation to describe the statement, "all the real numbers x within 2 units of 5."

Solution:

$$d(x,\ 5) = |x - 5|$$

Since $d(x,\ 5) \le 2$, we have $|x - 5| \le 2$.

19. Solve the equation $5x^4 - 12x^3 = 0$.

Solution:

$$5x^4 - 12x^3 = 0$$
$$x^3(5x - 12) = 0$$
$$x^3 = 0 \quad \text{or} \quad 5x - 12 = 0$$
$$x = 0 \quad \text{or} \quad x = \tfrac{12}{5}$$

23. Solve the equation $3\left(1 - \dfrac{1}{5t}\right) = 0$.

Solution:

$$3\left(1 - \frac{1}{5t}\right) = 0$$
$$1 - \frac{1}{5t} = 0$$
$$1 = \frac{1}{5t}$$
$$5t = 1$$
$$t = \frac{1}{5}$$

27. Solve the equation $4t^3 - 12t^2 + 8t = 0$.

Solution:

$$4t^3 - 12t^2 + 8t = 0$$
$$4t(t^2 - 3t + 2) = 0$$
$$4t(t - 1)(t - 2) = 0$$
$$t = 0, \ t = 1, \ t = 2$$

33. For the points $(2, 1)$ and $(14, 6)$, find (a) the distance between the two points, (b) the coordinates of the midpoint of the line segment between the two points, and (c) an equation of the circle whose diameter is the line segment between the two points.

Solution:

(a) $\ d = \sqrt{(14 - 2)^2 + (6 - 1)^2}$

$\qquad = \sqrt{144 + 25}$

$\qquad = \sqrt{169}$

$\qquad = 13$

(b) $\ m = \left(\dfrac{2 + 14}{2}, \ \dfrac{1 + 6}{2} \right)$

$\qquad = \left(8, \ \dfrac{7}{2} \right)$

(c) The length of the diameter is 13, so the length of the radius is $\frac{13}{2}$. The midpoint of the line segment is the center of the circle. Center: $\left(8, \frac{7}{2}\right)$ Radius: $\frac{13}{2}$

$$(x - 8)^2 + \left(y - \frac{7}{2} \right)^2 = \left(\frac{13}{2} \right)^2$$
$$x^2 - 16x + 64 + y^2 - 7y + \frac{49}{4} = \frac{169}{4}$$
$$x^2 + y^2 - 16x - 7y + 34 = 0$$

35. Find the intercepts of the graph of $2y^2 = x^3$ and check for symmetry with respect to each of the coordinate axes and the origin.

Solution:
The only intercept is the origin, $(0, 0)$.
The graph is symmetric with respect to the
x-axis since $2(-y)^2 = x^3$ results in
the original equation. Replacing x with
$-x$ or replacing both x and y with $-x$ and $-y$
does not yield equivalent equations. Thus,
the graph is not symmetric with respect to
either the y-axis or the origin.

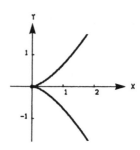

39. Find the intercepts of the graph of $y = x\sqrt{4 - x^2}$ and check for symmetry with respect to each of the coordinate axes and the origin.

Solution:
Let $y = 0$, then $0 = x\sqrt{4 - x^2}$ and $x = 0, \pm 2$.
x-intercepts: $(0, 0)$, $(2, 0)$, $(-2, 0)$
Let $x = 0$, then $y = 0\sqrt{4 - 0^2}$ and $y = 0$.
y-intercept: $(0, 0)$
The graph is symmetric with respect to the origin since

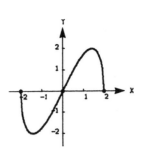

$$-y = -x\sqrt{4 - (-x)^2}$$
$$-y = -x\sqrt{4 - x^2}$$
$$y = x\sqrt{4 - x^2}$$

The graph is not symmetric with respect to either axis.

43. Determine the center and radius of the circle. Then, sketch the graph of $4x^2 + 4y^2 - 4x - 40y + 92 = 0$.

Solution:

$$4x^2 + 4y^2 - 4x - 40y + 92 = 0$$
$$x^2 + y^2 - x - 10y + 23 = 0$$
$$\left(x^2 - x + \tfrac{1}{4}\right) + \left(y^2 - 10y + 25\right) = -23 + \tfrac{1}{4} + 25$$
$$\left(x - \tfrac{1}{2}\right)^2 + (y - 5)^2 = \tfrac{9}{4}$$

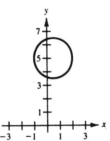

Center: $\left(\tfrac{1}{2}, 5\right)$ Radius: $\tfrac{3}{2}$

45. Sketch a graph of the equation $y - 2x - 3 = 0$.

Solution:
$y - 2x - 3 = 0$
x-intercept: $\left(-\tfrac{3}{2}, 0\right)$
y-intercept: $(0, 3)$

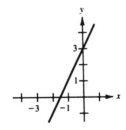

51. Sketch a graph of the equation $y = \sqrt{25 - x^2}$.

Solution:

$y = \sqrt{25 - x^2}$

x-intercepts: $(5, 0)$, $(-5, 0)$

y-intercept: $(0, 5)$

y-axis symmetry

Domain: $[-5, 5]$

Range: $[0, 5]$

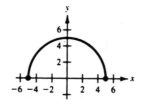

53. Sketch a graph of the equation $y = \frac{1}{4}(x + 1)^3$.

Solution:

$y = \frac{1}{4}(x + 1)^3$

x-intercept: $(-1, 0)$

y-intercept: $(0, \frac{1}{4})$

Domain: $(-\infty, \infty)$

Range: $(-\infty, \infty)$

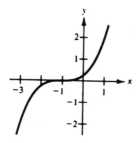

59. Determine the domain of the function

$$g(s) = \frac{5}{3s - 9}.$$

Solution:

The domain of $g(s) = 5/(3s - 9)$ includes all real numbers except $s = 3$, since this value would yield a zero in the denominator.

63. For $f(x) = \sqrt{x + 1}$ (a) find f^{-1}, (b) sketch the graphs of f and f^{-1} on the same coordinate plane, and (c) verify that $f^{-1}(f(x)) = x = f(f^{-1}(x))$.

Solution:

(a)
$$y = \sqrt{x + 1}, \quad y \geq 0$$
$$y^2 = x + 1$$
$$x = y^2 - 1$$
$$f^{-1}(x) = x^2 - 1, \quad x \geq 0$$

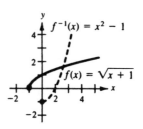

(c) $f^{-1}[f(x)] = f^{-1}(\sqrt{x + 1})$
$$= (\sqrt{x + 1})^2 - 1 = (x + 1) - 1 = x$$
$$f[f^{-1}(x)] = f(x^2 - 1)$$
$$= \sqrt{(x^2 - 1) + 1}$$
$$= \sqrt{x^2} = x, \quad x \geq 0$$

67. Restrict the domain of the function $f(x) = 2(x-4)^2$ to an interval where the function is increasing and determine f^{-1} over that interval.

Solution:
$f(x) = 2(x-4)^2$ is increasing on the interval $[4, \infty)$. It is decreasing on the interval $(-\infty, 4)$.

$$f(x) = 2(x-4)^2, \ x \geq 4$$
$$y = 2(x-4)^2, \ x \geq 4$$
$$\sqrt{y} = \sqrt{2}(x-4)$$
$$\sqrt{y/2} = x - 4$$
$$x = \sqrt{y/2} + 4$$
$$f^{-1}(x) = \sqrt{x/2} + 4$$

73. Let $f(x) = 3 - 2x$, $\ g(x) = \sqrt{x}$, and $h(x) = 3x^2 + 2$. Find $(h \circ g)(7)$.

Solution:

$$(h \circ g)(7) = h(g(7))$$
$$= h(\sqrt{7})$$
$$= 3(\sqrt{7})^2 + 2$$
$$= 23$$

77. A wire 24 inches long is to be cut into four pieces to form a rectangle whose shortest side has a length of x. Express the area A of the rectangle as a function of x. Determine the domain of the function and sketch its graph over that domain.

Solution:
Let y be the longer side of the rectangle. Then we have $A = xy$. Since the perimeter is 24 inches, we have $2x + 2y = 24$ or $y = (24 - 2x)/2 = 12 - x$. The area equation now becomes: $A = xy = x(12 - x)$. To find the domain of A, we realize that area is a nonnegative quantity. Thus, $x(12 - x) \geq 0$. This gives us the interval $[0, 12]$. We also have the further restriction that x is the shortest side. This occurs on the interval $[0, 6]$.

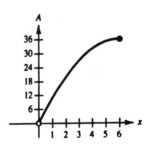

79. A group of farmers agree to share equally in the cost of a $48,000 piece of machinery. If they could find two more farmers to join the group, each person's share of the cost would decrease by $4000. How many farmers are presently in the group?

Solution:

Let x = number of farmers in the group

$$\text{Cost per farmer} = \frac{48,000}{x}$$

If two more farmers join the group, the cost per farmer will be $\dfrac{48,000}{x+2}$.

Since this new cost is $4000 less than the original cost,

$$\frac{48,000}{x} - 4000 = \frac{48,000}{x+2}$$

$$48,000(x+2) - 4000x(x+2) = 48,000x$$

$$12(x+2) - x(x+2) = 12x$$

$$12x + 24 - x^2 - 2x = 12x$$

$$0 = x^2 + 2x - 24$$

$$0 = (x+6)(x-4)$$

$$x = -6, \text{ extraneous} \quad \text{OR} \quad x = 4$$

$$x = 4 \text{ farmers}$$

Practice Test for Chapter 1

Solutions to Selected Exercises

1. Evaluate $-|-17|-17$.

2. Find the distance between a and b; $a = \frac{12}{5}$, $b = -\frac{7}{15}$.

3. Use absolute value notation to describe the expression "The distance between z and -5 is no more than 12."

4. Use absolute value notation to define the given interval on the real line.

5. Solve the equation $\dfrac{6x+5}{2x-9} = \dfrac{3}{5}$.

6. Use the Quadratic Formula to solve the equation $4x^2 + 3x - 5 = 0$.

7. Use a calculator to solve the equation $21.4x^2 + 6.9x - 1.4 = 0$. Round your answers to three decimal places.

8. Find all real solutions to the equation $x^6 - 7x^3 - 8 = 0$.

9. Find the distance between the points $(4, 7)$ and $(-2, 5)$.

10. Find the midpoint of the line segment joining $(-1, 16)$ and $(3, -5)$.

11. Find x so that the distance between the points $(2, 0)$ and $(-6, x)$ is 9.

12. Find the x- and y-intercepts of the graph of the equation $y = x\sqrt{3-x}$.

13. Check for symmetry with respect to both axes and the origin: $y = \dfrac{x^2}{x^3 - 1}$.

14. Graph $y = \sqrt{x+2}$.

15. Graph $y = |x - 3|$.

16. Write the equation of the circle in standard form: $x^2 + y^2 - 14x + 6y + 42 = 0$.

17. Given $f(x) = 5x + 11$, find $\dfrac{f(x) - f(2)}{x - 2}$.

18. Find the domain of $f(x) = \sqrt{\dfrac{x-1}{x+3}}$.

19. An open box is to be made from a rectangular piece of cardboard, 16 inches by 11 inches, by cutting equal squares from each corner and turning up the sides. Write an equation for the volume V of the box as a function of its height x.

In Exercises 20–22, sketch the graph of the given function.

20. $f(x) = x^2 - 9$

21. $f(x) = -1 + |x|$

22. $f(x) = \begin{cases} 3x + 2, & x \geq 1 \\ x^2 - 1, & x < 1 \end{cases}$

23. Given $f(x) = x^2 + 2$ and $g(x) = 3x - 8$, find $(f \circ g)(x)$.

24. Find the inverse of $f(x) = \dfrac{x+3}{x}$.

25. Given $f(x) = \dfrac{x^3}{4}$ and $g(x) = 3x$, find $g^{-1} \circ f^{-1}$.

CHAPTER 2

Trigonometry

SECTION 2.1

Radian and Degree Measure

- If two angles have the same initial and terminal sides, they are coterminal angles.

- The radian measure of a central angle θ is found by taking the arc length s and dividing it by the radius r.

$$\theta = \frac{s}{r}$$

- You should know the following about angles:

  (a) θ is acute if $0 < \theta < \pi/2$.

  (b) θ is a right angle if $\theta = \pi/2$.

  (c) θ is obtuse if $\pi/2 < \theta < \pi$.

  (d) α and β are complementary if $\alpha + \beta = \pi/2$.

  (e) α and β are supplementary if $\alpha + \beta = \pi$.

- To convert degrees to radians, multiply by $\pi/180$.

- To convert radians to degrees, multiply by $180/\pi$.

- You should be able to convert angles to degrees, minutes, and seconds.

  (a) One minute: $1' = \dfrac{1}{60}(1°)$

  (b) One second: $1'' = \dfrac{1}{60}(1') = \dfrac{1}{3600}(1°)$

- $\text{Speed} = \dfrac{\text{distance}}{\text{time}} = \dfrac{s}{t}$

- $\text{Angular speed} = \dfrac{\theta}{t}$

Solutions to Selected Exercises

3. Determine the quadrant in which the following angles lie. (The angle measure is given in radians.)

(a) $-\dfrac{\pi}{12}$

(b) $-\dfrac{11\pi}{9}$

Solution:

(a) $-\dfrac{\pi}{12}$ lies in Quadrant IV.

(b) $-\dfrac{11\pi}{9}$ lies in Quadrant II.

9. Determine the quadrant in which the following angles lie.

(a) $-132°\,50'$

(b) $-336°$

Solution:

(a) $-132°\,50'$ lies in Quadrant III.

(b) $-336°$ lies in Quadrant I.

13. Sketch the given angle in standard position.

(a) $-\dfrac{7\pi}{4}$

(b) $-\dfrac{5\pi}{2}$

Solution:

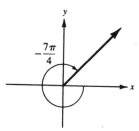

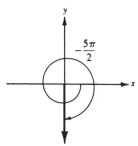

17. Determine two coterminal angles (one positive and one negative) for (a) $\theta = \pi/9$ and (b) $\theta = 4\pi/3$. Give the answers in radians.

Solution:

(a) Coterminal angles for $\dfrac{\pi}{9}$:

$$2\pi + \frac{\pi}{9} = \frac{19\pi}{9}$$

$$\frac{\pi}{9} - 2\pi = -\frac{17\pi}{9}$$

(b) Coterminal angles for $\dfrac{4\pi}{3}$:

$$2\pi + \frac{4\pi}{3} = \frac{10\pi}{3}$$

$$\frac{4\pi}{3} - 2\pi = -\frac{2\pi}{3}$$

21. Determine two coterminal angles (one positive and one negative) for (a) $\theta = 36°$ and (b) $\theta = -45°$. Give the answers in degrees.

Solution:

(a) Coterminal angles for $36°$:

$360° + 36° = 396°$

$36° - 360° = -324°$

(b) Coterminal angles for $-45°$:

$360° + (-45°) = 315°$

$-45° - 360° = -405°$

27. Express (a) $-20°$ and (b) $-240°$ in radian measure as a multiple of π. (Do not use a calculator.)

Solution:

(a) $-20° = -20\left(\dfrac{\pi}{180}\right) = -\dfrac{\pi}{9}$

(b) $-240° = -240\left(\dfrac{\pi}{180}\right) = -\dfrac{4\pi}{3}$

31. Express (a) $7\pi/3$ and (b) $-11\pi/30$ in degree measure. (Do not use a calculator.)

Solution:

(a) $\dfrac{7\pi}{3} = \dfrac{7\pi}{3}\left(\dfrac{180}{\pi}\right) = 420°$

(b) $-\dfrac{11\pi}{30} = -\dfrac{11\pi}{30}\left(\dfrac{180}{\pi}\right) = -66°$

35. Convert $-216.35°$ from degrees to radian measure. List your answer to three decimal places.

Solution:

$$-216.35° = -216.35\left(\dfrac{\pi}{180}\right) \approx -3.776 \text{ radians}$$

39. Convert $-0.83°$ from degrees to radian measure. List your answer to three decimal places.

Solution:

$$-0.83° = -0.83\left(\dfrac{\pi}{180}\right) \approx -0.014 \text{ radians}$$

41. Convert $\pi/7$ from radian to degree measure. List your answer to three decimal places.

Solution:

$$\dfrac{\pi}{7} = \dfrac{\pi}{7}\left(\dfrac{180}{\pi}\right) \approx 25.714°$$

45. Convert -4.2π from radian to degree measure. List your answer to three decimal places.

Solution:

$$-4.2\pi = -4.2\pi\left(\dfrac{180}{\pi}\right) = -756°$$

51. Convert (a) $85° \, 18' \, 30''$ and (b) $330° \, 25''$ to decimal form.

Solution:

(a) $85° \, 18' \, 30'' = 85 + \frac{18}{60} + \frac{30}{3600} \approx 85.308°$ (b) $330° \, 25'' = 330 + \frac{25}{3600} \approx 330.007°$

53. Convert (a) $240.6°$ and (b) $-145.8°$ to $D° \, M' \, S''$ form.

Solution:

(a) $240.6° = 240 + 0.6(60) = 240° \, 36'$ (b) $-145.8° = -[145 + 0.8(60)] = -145° \, 48'$

57. Find the radian measure of the central angle using radius $r = 10$ inches and arc length $s = 4$ inches.

Solution:

$$s = r\theta \Rightarrow \theta = \frac{s}{r}$$

$$\theta = \frac{4 \text{ inches}}{10 \text{ inches}} = \frac{2}{5} \text{ radians}$$

63. Find the length of the arc on the circle of radius $r = 6$ meters subtended by the central angle $\theta = 2$ radians.

Solution:

$$s = r\theta$$

$$s = (6)(2) = 12 \text{ m}$$

69. Assuming that the earth is a sphere of radius 4000 miles, what is the difference in latitude of two cities, one of which is 325 miles due north of the other?

Solution:
Use the formula $\theta = s/r$.

$$\theta = \frac{325}{4000} \cdot \frac{180}{\pi} = 4.655°$$

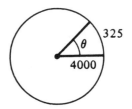

73. A car is moving at the rate of 50 miles per hour, and the diameter of each of its wheels is 2.5 feet.

(a) Find the number of revolutions per minute that the wheels are rotating.
(b) Find the *angular speed* of the wheels.

Solution:
(a) 50 miles per hour $= 50(5280)/60 = 4400$ feet per minute
The circumference of the tire is $C = 2.5\pi$ feet.
The number of revolutions per minute is $r = 4400/2.5\pi \approx 560.2$

(b) The angular speed is θ/t.

$$\theta = \frac{4400}{2.5\pi}(2\pi) = 3520 \text{ radians}$$

$$\text{Angular speed} = \frac{3520 \text{ radians}}{1 \text{ minute}} = 3520 \text{ rad/min}$$

SECTION 2.2

Trigonometric Functions and the Unit Circle

■ You should know the definition of the trigonometric functions of t.

(a) $\sin t = y$
(b) $\cos t = x$
(c) $\tan t = y/x, \quad x \neq 0$
(d) $\cot t = x/y, \quad y \neq 0$
(e) $\sec t = 1/x, \quad x \neq 0$
(f) $\csc t = 1/y, \quad y \neq 0$

■ The cosine and secant functions are even.

(a) $\cos(-t) = \cos t$
(b) $\sec(-t) = \sec t$

■ The other four trigonometric functions are odd.

(a) $\sin(-t) = -\sin t$
(b) $\tan(-t) = -\tan t$
(c) $\cot(-t) = -\cot t$
(d) $\csc(-t) = -\csc t$

■ You should be able to evaluate trigonometric functions with a calculator.

Solutions to Selected Exercises

3. Find the point (x, y) on the unit circle that corresponds to $t = 5\pi/6$.

Solution:
Using Figure 5.19 in the text, move counterclockwise to obtain the second quadrant point

$$(x, y) = \left(-\frac{\sqrt{3}}{2}, \frac{1}{2}\right).$$

5. Find the point (x, y) on the unit circle that corresponds to $\theta = 4\pi/3$.

Solution:
Using Figure 5.19 again and the fact that $4\pi/3 = 8\pi/6$, we obtain the point

$$(x, y) = \left(-\frac{1}{2}, -\frac{\sqrt{3}}{2}\right).$$

11. Evaluate the given trigonometric function.

(a) $\sin\left(-\dfrac{\pi}{6}\right)$ (b) $\cos\left(-\dfrac{\pi}{6}\right)$ (c) $\tan\left(-\dfrac{\pi}{6}\right)$

Solution:

$t = \dfrac{\pi}{6}$ corresponds to the point $\left(\dfrac{\sqrt{3}}{2},\ \dfrac{1}{2}\right)$.

(a) $\sin\left(-\dfrac{\pi}{6}\right) = -\sin\dfrac{\pi}{6} = -y = -\dfrac{1}{2}$

(b) $\cos\left(-\dfrac{\pi}{6}\right) = \cos\dfrac{\pi}{6} = x = \dfrac{\sqrt{3}}{2}$

(c) $\tan\left(-\dfrac{\pi}{6}\right) = -\tan\dfrac{\pi}{6} = -\dfrac{y}{x} = -\dfrac{1/2}{\sqrt{3}/2} = -\dfrac{1}{\sqrt{3}} = -\dfrac{\sqrt{3}}{3}$

17. Evaluate the given trigonometric function.

(a) $\sin\dfrac{11\pi}{6}$ (b) $\cos\dfrac{11\pi}{6}$ (c) $\tan\dfrac{11\pi}{6}$

Solution:

$t = \dfrac{11\pi}{6}$ corresponds to the point $\left(\dfrac{\sqrt{3}}{2},\ -\dfrac{1}{2}\right)$.

(a) $\sin\dfrac{11\pi}{6} = y = -\dfrac{1}{2}$

(b) $\cos\dfrac{11\pi}{6} = x = \dfrac{\sqrt{3}}{2}$

(c) $\tan\dfrac{11\pi}{6} = \dfrac{y}{x} = \dfrac{-1/2}{\sqrt{3}/2} = -\dfrac{1}{\sqrt{3}} = -\dfrac{\sqrt{3}}{3}$

21. Evaluate the six trigonometric functions for $t = \pi/4$.

Solution:

$t = \dfrac{\pi}{4}$ corresponds to the point $\left(\dfrac{\sqrt{2}}{2},\ \dfrac{\sqrt{2}}{2}\right)$.

$\sin\dfrac{\pi}{4} = y = \dfrac{\sqrt{2}}{2}$

$\cos\dfrac{\pi}{4} = x = \dfrac{\sqrt{2}}{2}$

$\tan\dfrac{\pi}{4} = \dfrac{y}{x} = 1$

$\cot\dfrac{\pi}{4} = \dfrac{x}{y} = 1$

$\sec\dfrac{\pi}{4} = \dfrac{1}{x} = \sqrt{2}$

$\csc\dfrac{\pi}{4} = \dfrac{1}{y} = \sqrt{2}$

27. Use the periodic nature of the sine and cosine to evaluate $\sin 3\pi$.

Solution:

$$\sin 3\pi = \sin(2\pi + \pi) = \sin \pi = 0$$

31. Use the periodic nature of the sine and cosine to evaluate $\cos \dfrac{19\pi}{6}$.

Solution:

$$\cos \frac{19\pi}{6} = \cos\left(2\pi + \frac{7\pi}{6}\right) = \cos \frac{7\pi}{6} = -\frac{\sqrt{3}}{2}$$

35. Use $\sin t = \frac{1}{3}$ to evaluate the indicated functions.

    (a) $\sin(-t)$ (b) $\csc(-t)$

Solution:

    (a) $\sin(-t) = -\sin t = -\dfrac{1}{3}$

    (b) $\csc(-t) = -\csc t = -\dfrac{1}{\sin t} = -\dfrac{1}{1/3} = -3$

41. Use a calculator to evaluate $\cos(-3)$. [Set your calculator in radian mode and round your answer to four decimal places.]

Solution:

$$\cos(-3) = \cos 3 \approx -0.9899925 \approx -0.9900$$

43. Use a calculator to evaluate $\tan(\pi/10)$. [Set your calculator in radian mode and round your answer to four decimal places.]

Solution:

$$\tan \frac{\pi}{10} \approx 0.3249197 \approx 0.3249$$

47. Use a calculator to evaluate $\csc 0.8$. [Set your calculator in radian mode and round your answer to four decimal places.]

Solution:

$$\csc 0.8 = \frac{1}{\sin 0.8} \approx 1.3940078 \approx 1.3940$$

SECTION 2.3

Trigonometric Functions of an Acute Angle

■ You should know the right triangle definition of trigonometric functions.

(a) $\sin \theta = \dfrac{\text{opp}}{\text{hyp}}$

(b) $\cos \theta = \dfrac{\text{adj}}{\text{hyp}}$

(c) $\tan \theta = \dfrac{\text{opp}}{\text{adj}}$

(d) $\cot \theta = \dfrac{\text{adj}}{\text{opp}}$

(e) $\sec \theta = \dfrac{\text{hyp}}{\text{adj}}$

(f) $\csc \theta = \dfrac{\text{hyp}}{\text{opp}}$

■ You should know the sine, cosine, and tangent of the special angles 30°, 45°, and 60°.

(a) For 45°, use the triangle

(b) For 30° and 60°, use the triangle

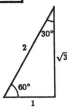

■ You should know the fundamental trigonometric identities.

(a) $\sin \theta = \dfrac{1}{\csc \theta}$

(b) $\cos \theta = \dfrac{1}{\sec \theta}$

(c) $\tan \theta = \dfrac{1}{\cot \theta}$

(d) $\cot \theta = \dfrac{1}{\tan \theta}$

(e) $\sec \theta = \dfrac{1}{\cos \theta}$

(f) $\csc \theta = \dfrac{1}{\sin \theta}$

(g) $\tan \theta = \dfrac{\sin \theta}{\cos \theta}$

(h) $\cot \theta = \dfrac{\cos \theta}{\sin \theta}$

(i) $\sin^2 \theta + \cos^2 \theta = 1$

(j) $1 + \tan^2 \theta = \sec^2 \theta$

(k) $1 + \cot^2 \theta = \csc^2 \theta$

Solutions to Selected Exercises

3. Find the exact value of the six trigonometric functions of the angle θ given in the figure. (Use the Pythagorean Theorem to find the third side of the triangle.)

Solution:

$$b = \sqrt{c^2 - a^2}$$
$$b = \sqrt{(5)^2 - (4)^2} = \sqrt{9} = 3$$

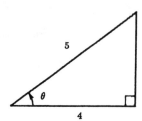

$$\sin\theta = \tfrac{3}{5} \qquad \csc\theta = \tfrac{5}{3}$$
$$\cos\theta = \tfrac{4}{5} \qquad \sec\theta = \tfrac{5}{4}$$
$$\tan\theta = \tfrac{3}{4} \qquad \cot\theta = \tfrac{4}{3}$$

9. Sketch a right triangle corresponding to $\sin\theta = \tfrac{2}{3}$, and find the other five trigonometric functions of θ.

Solution:

$$\sin\theta = \frac{2}{3}$$
$$\text{Adjacent side} = \sqrt{9 - 4} = \sqrt{5}$$

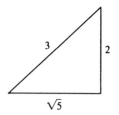

$$\cos\theta = \frac{\sqrt{5}}{3} \qquad \sec\theta = \frac{3}{\sqrt{5}} = \frac{3\sqrt{5}}{5}$$
$$\tan\theta = \frac{2}{\sqrt{5}} = \frac{2\sqrt{5}}{5} \qquad \csc\theta = \frac{3}{2}$$
$$\cot\theta = \frac{\sqrt{5}}{2}$$

13. Sketch a right triangle corresponding to $\tan\theta = 3$, and find the other five trigonometric functions of θ.

Solution:

$$\tan\theta = 3 = \frac{3}{1}$$
$$\text{Hypotenuse} = \sqrt{1 + 9} = \sqrt{10}$$

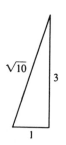

$$\sin\theta = \frac{3}{\sqrt{10}} = \frac{3\sqrt{10}}{10} \qquad \sec\theta = \sqrt{10}$$
$$\cos\theta = \frac{1}{\sqrt{10}} = \frac{\sqrt{10}}{10} \qquad \csc\theta = \frac{\sqrt{10}}{3}$$
$$\cot\theta = \frac{1}{3}$$

19. Use $\csc\theta = 3$ and $\sec\theta = \frac{3\sqrt{2}}{4}$ to find the following trigonometric functions.

(a) $\sin\theta$ (b) $\cos\theta$

(c) $\tan\theta$ (d) $\sec(90° - \theta)$

Solution:

(a) $\sin \theta = \dfrac{1}{\csc \theta} = \dfrac{1}{3}$

(b) $\cos \theta = \dfrac{1}{\sec \theta} = \dfrac{4}{3\sqrt{2}} = \dfrac{2\sqrt{2}}{3}$

(c) $\tan \theta = \dfrac{\sin \theta}{\cos \theta} = \dfrac{1/3}{(2\sqrt{2})/3} = \dfrac{1}{2\sqrt{2}} = \dfrac{\sqrt{2}}{4}$

(d) $\sec(90° - \theta) = \csc \theta = 3$

21. Evaluate (a) $\cos 60°$ and (b) $\tan 30°$ by memory or by constructing an appropriate triangle.

Solution:

(a) $\cos 60° = \dfrac{1}{2}$

(b) $\tan 30° = \dfrac{1}{\sqrt{3}} = \dfrac{\sqrt{3}}{3}$

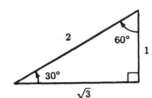

25. Use a calculator to evaluate (a) $\sin 10°$ and (b) $\cos 80°$. Round your answers to four decimal places. (Be sure the calculator is in the correct mode.)

Solution:

(a) $\sin 10° \approx 0.1736$

(b) $\cos 80° = \sin 10° \approx 0.1736$

31. Use a calculator to evaluate (a) $\cot(\pi/16)$ and (b) $\tan(\pi/16)$. Round your answers to four decimal places.

Solution:

Make sure that your calculator is in radian mode.

(a) $\cot \dfrac{\pi}{16} = \dfrac{1}{\tan(\pi/16)} \approx 5.0273$

(b) $\tan \dfrac{\pi}{16} \approx 0.1989$

35. Find the value of θ in degrees $(0° < \theta < 90°)$ and radians $(0 < \theta < \pi/2)$ for (a) $\sin \theta = 1/2$ and (b) $\csc \theta = 2$ without a calculator.

Solution:

(a) $\sin \theta = \dfrac{1}{2}$

$\theta = 30° = \dfrac{\pi}{6}$

(b) $\csc \theta = 2 \Rightarrow \sin \theta = \dfrac{1}{2}$ Same as (a)

$\theta = 30° = \dfrac{\pi}{6}$

39. Find the value of θ in degrees $(0° < \theta < 90°)$ and radians $(0 < \theta < \pi/2)$ for (a) $\csc\theta = (2\sqrt{3})/3$ and (b) $\sin\theta = \sqrt{2}/2$ without a calculator.

Solution:

(a) $\csc\theta = \dfrac{2\sqrt{3}}{3} = \dfrac{2}{\sqrt{3}}$

$\quad\quad\theta = 60° = \dfrac{\pi}{3}$

(b) $\sin\theta = \dfrac{\sqrt{2}}{2} = \dfrac{1}{\sqrt{2}}$

$\quad\quad\theta = 45° = \dfrac{\pi}{4}$

47. Solve for x.

Solution:

$$\tan 60° = \frac{25}{x}$$

$$\sqrt{3} = \frac{25}{x}$$

$$x = \frac{25}{\sqrt{3}} = \frac{25\sqrt{3}}{3}$$

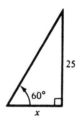

51. Solve for y.

Solution:

$$\sin 50° = \frac{y}{12}$$

$$0.7660 = \frac{y}{12}$$

$$y = 9.19$$

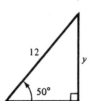

53. A six-foot person standing 12 feet from a streetlight casts an eight-foot shadow, as shown in the figure. What is the height of the streetlight?

Solution:

$$\frac{h}{6} = \frac{20}{8}$$

$$h = \frac{120}{8} = 15 \text{ ft}$$

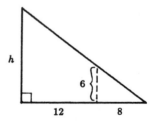

57. From a 150-foot observation tower on the coast, a Coast Guard officer sights a boat in difficulty. The angle of depression of the boat is 4°, as shown in the figure. How far is the boat from the shoreline?

Solution:

Let x = distance from the boat to the shoreline.

$$\tan 4° = \frac{150}{x}$$

$$x = \frac{150}{\tan 4°} \approx 2145.10 \text{ ft}$$

61. Determine whether the statement is true or false, and give reasons.

$$\sin 45° + \cos 45° = 1.$$

Solution:

False; $\sin 45° + \cos 45° = \dfrac{\sqrt{2}}{2} + \dfrac{\sqrt{2}}{2} = \sqrt{2} \neq 1$

SECTION 2.4

Trigonometric Functions of Any Angle

- You should know the trigonometric functions of any angle θ in standard position with (x, y) on the terminal side of θ and $r = \sqrt{x^2 + y^2}$.

  (a) $\sin \theta = \dfrac{y}{r}$ (b) $\cos \theta = \dfrac{x}{r}$

  (c) $\tan \theta = \dfrac{y}{x}, \quad x \neq 0$ (d) $\cot \theta = \dfrac{x}{y}, \quad y \neq 0$

  (e) $\sec \theta = \dfrac{r}{x}, \quad x \neq 0$ (f) $\csc \theta = \dfrac{r}{y}, \quad y \neq 0$

- You should know the signs of the trigonometric functions in the four quadrants.

- You should be able to find the trigonometric functions of the quadrant angles (if they exist).

  (a) For 0, use $(1, 0)$.

  (b) For $\pi/2$, use $(0, 1)$.

  (c) For π, use $(-1, 0)$.

  (d) For $3\pi/2$, use $(0, -1)$.

- You should be able to use reference angles with the special angles to find trigonometric values.

Solutions to Selected Exercises

3. Determine the exact value of the six trigonometric functions of the given angle θ.

(a)

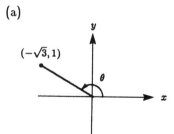

(b)

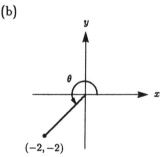

Solution:

(a) $x = -\sqrt{3}$, $y = 1$, $r = \sqrt{3 + 1} = 2$

$$\sin \theta = \frac{y}{r} = \frac{1}{2} \qquad\qquad \csc \theta = \frac{r}{y} = 2$$

$$\cos \theta = \frac{x}{r} = -\frac{\sqrt{3}}{2} \qquad\qquad \sec \theta = \frac{r}{x} = -\frac{2\sqrt{3}}{3}$$

$$\tan \theta = \frac{y}{x} = -\frac{1}{\sqrt{3}} = -\frac{\sqrt{3}}{3} \qquad\qquad \cot \theta = \frac{x}{y} = -\sqrt{3}$$

(b) $x = -2$, $y = -2$, $r = \sqrt{4 + 4} = 2\sqrt{2}$

$$\sin \theta = \frac{y}{r} = -\frac{2}{2\sqrt{2}} = -\frac{\sqrt{2}}{2} \qquad\qquad \csc \theta = \frac{r}{y} = \frac{2\sqrt{2}}{-2} = -\sqrt{2}$$

$$\cos \theta = \frac{x}{r} = -\frac{2}{2\sqrt{2}} = -\frac{\sqrt{2}}{2} \qquad\qquad \sec \theta = \frac{r}{x} = \frac{2\sqrt{2}}{-2} = -\sqrt{2}$$

$$\tan \theta = \frac{y}{x} = \frac{-2}{-2} = 1 \qquad\qquad \cot \theta = \frac{x}{y} = \frac{-2}{-2} = 1$$

7. The given point is on the terminal side of an angle in standard position. Determine the exact value of the six trigonometric functions of the angle.

(a) $(-4, \ 10)$ \qquad\qquad\qquad (b) $(3, \ -5)$

Solution:

(a) $x = -4$, $y = 10$, $r = \sqrt{16 + 100} = 2\sqrt{29}$

$$\sin \theta = \frac{10}{2\sqrt{29}} = \frac{5\sqrt{29}}{29} \qquad\qquad \csc \theta = \frac{2\sqrt{29}}{10} = \frac{\sqrt{29}}{5}$$

$$\cos \theta = \frac{-4}{2\sqrt{29}} = -\frac{2\sqrt{29}}{29} \qquad\qquad \sec \theta = \frac{2\sqrt{29}}{-4} = -\frac{\sqrt{29}}{2}$$

$$\tan \theta = \frac{10}{-4} = -\frac{5}{2} \qquad\qquad \cot \theta = \frac{-4}{10} = -\frac{2}{5}$$

(b) $x = 3$, $y = -5$, $r = \sqrt{9 + 25} = \sqrt{34}$

$$\sin \theta = \frac{-5}{\sqrt{34}} = -\frac{5\sqrt{34}}{34} \qquad\qquad \csc \theta = -\frac{\sqrt{34}}{5}$$

$$\cos \theta = \frac{3}{\sqrt{34}} = \frac{3\sqrt{34}}{34} \qquad\qquad \sec \theta = \frac{\sqrt{34}}{3}$$

$$\tan \theta = -\frac{5}{3} \qquad\qquad \cot \theta = -\frac{3}{5}$$

9. Use the two similar triangles in the figure to find (a) the unknown sides of the triangles and (b) the six trigonometric functions of the angles α_1 and α_2.

Solution:

Given $a_1 = 3$, $b_1 = 4$, $a_2 = 9$,

(a) $c_1 = \sqrt{a_1{}^2 + b_1{}^2} = \sqrt{9 + 16} = 5$

$$\frac{a_1}{a_2} = \frac{b_1}{b_2}$$

$$\frac{3}{9} = \frac{4}{b_2}$$

$b_2 = 12$

$c_2 = \sqrt{a_2{}^2 + b_2{}^2} = \sqrt{81 + 144} = 15$

(b)

$\sin \alpha_1 = \sin \alpha_2 = \dfrac{3}{5}$	$\csc \alpha_1 = \csc \alpha_2 = \dfrac{5}{3}$
$\cos \alpha_1 = \cos \alpha_2 = \dfrac{4}{5}$	$\sec \alpha_1 = \sec \alpha_2 = \dfrac{5}{4}$
$\tan \alpha_1 = \tan \alpha_2 = \dfrac{3}{4}$	$\cot \alpha_1 = \cot \alpha_2 = \dfrac{4}{3}$

13. Determine the quadrant in which θ lies.

(a) $\sin \theta < 0$ and $\cos \theta < 0$
(b) $\sin \theta > 0$ and $\cos \theta < 0$

Solution:

(a) $\sin \theta < 0 \Rightarrow \theta$ lies in Quadrant III or in Quadrant IV.
$\cos \theta < 0 \Rightarrow \theta$ lies in Quadrant II or in Quadrant III.
$\sin \theta < 0$ *and* $\cos \theta < 0 \Rightarrow \theta$ lies in Quadrant III.

(b) $\sin \theta > 0 \Rightarrow \theta$ lies in Quadrant I or in Quadrant II.
$\cos \theta < 0 \Rightarrow \theta$ lies in Quadrant II or in Quadrant III.
$\sin \theta > 0$ *and* $\cos \theta < 0 \Rightarrow \theta$ lies in Quadrant II.

17. Find the exact value of the six trigonometric functions of θ, given θ lies in Quadrant II and $\sin \theta = \frac{3}{5}$.

Solution:

$y = 3$, $r = 5$, $x = -\sqrt{25 - 9} = -4$, x is negative since θ lies in Quadrant II.

$\sin \theta = \frac{3}{5}$	$\csc \theta = \frac{5}{3}$
$\cos \theta = -\frac{4}{5}$	$\sec \theta = -\frac{5}{4}$
$\tan \theta = -\frac{3}{4}$	$\cot \theta = -\frac{4}{3}$

21. Find the exact value of the six trigonometric functions of θ, given $\sin \theta > 0$ and $\sec \theta = -2$.

Solution:

θ is in Quadrant II.

$\sec \theta = \dfrac{2}{-1}, \ r = 2, \ x = -1, \ y = \sqrt{4-1} = \sqrt{3}$

$$\sin \theta = \frac{\sqrt{3}}{2} \qquad \csc \theta = \frac{2\sqrt{3}}{3}$$

$$\cos \theta = -\frac{1}{2} \qquad \sec \theta = -2$$

$$\tan \theta = -\sqrt{3} \qquad \cot \theta = -\frac{\sqrt{3}}{3}$$

25. Find the exact value of the six trigonometric functions of θ, given the terminal side of θ is in Quadrant III and lies on the line $y = 2x$.

Solution:

To find a point on the terminal side of θ, use any point on the line $y = 2x$ that lies in Quadrant III. $(-1, -2)$ is one such point.

$x = -1, \ y = -2, \ r = \sqrt{5}$

$$\sin \theta = -\frac{2}{\sqrt{5}} = -\frac{2\sqrt{5}}{5} \qquad \csc \theta = \frac{\sqrt{5}}{-2} = -\frac{\sqrt{5}}{2}$$

$$\cos \theta = -\frac{1}{\sqrt{5}} = -\frac{\sqrt{5}}{5} \qquad \sec \theta = \frac{\sqrt{5}}{-1} = -\sqrt{5}$$

$$\tan \theta = \frac{-2}{-1} = 2 \qquad \cot \theta = \frac{-1}{-2} = \frac{1}{2}$$

29. Find the reference angle θ', and draw a sketch for (a) $\theta = -245°$ and (b) $\theta = -72°$.

Solution:

(a) $\theta = -245°$

     $\theta' = 245° - 180°$

     $\theta' = 65°$

(b) $\theta = -72°$

     $\theta' = |-72°| = 72°$

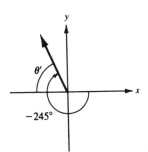

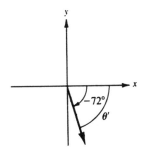

33. Find the reference angle θ' and draw a sketch for (a) $\theta = 3.5$ and (b) $\theta = 5.8$.

Solution:

(a) $\theta = 3.5$

       $\theta' = 3.5 - \pi$

       $\theta' \approx 0.3584$

(b) $\theta = 5.8$

       $\theta' = 2\pi - 5.8$

       $\theta' \approx 0.4832$

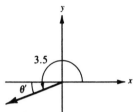
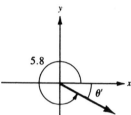

35. Evaluate the sine, cosine, and tangent of the angles without using a calculator.

(a) $225°$

(b) $-225°$

Solution:

(a) The reference angle of $225°$ is $45°$, and $225°$ lies in Quadrant III.

$$\sin 225° = -\sin 45° = -\frac{\sqrt{2}}{2}$$

$$\cos 225° = -\cos 45° = -\frac{\sqrt{2}}{2}$$

$$\tan 225° = \tan 45° = 1$$

(b) The reference angle is $45°$, and $-225°$ lies in Quadrant II.

$$\sin(-225°) = \sin 45° = \frac{\sqrt{2}}{2}$$

$$\cos(-225°) = -\cos 45° = -\frac{\sqrt{2}}{2}$$

$$\tan(-225°) = -\tan 45° = -1$$

39. Evaluate the sine, cosine, and tangent of the angles without using a calculator.

(a) $\dfrac{4\pi}{3}$

(b) $\dfrac{2\pi}{3}$

Solution:

(a) The reference angle of $4\pi/3$ is $\pi/3$, and $4\pi/3$ lies in Quadrant III.

$$\sin \frac{4\pi}{3} = -\sin \frac{\pi}{3} = -\frac{\sqrt{3}}{2}$$

$$\cos \frac{4\pi}{3} = -\cos \frac{\pi}{3} = -\frac{1}{2}$$

$$\tan \frac{4\pi}{3} = \tan \frac{\pi}{3} = \sqrt{3}$$

(b) The reference angle of $2\pi/3$ is $\pi/3$, and $2\pi/3$ lies in Quadrant II.

$$\sin\frac{2\pi}{3} = \sin\frac{\pi}{3} = \frac{\sqrt{3}}{2}$$

$$\cos\frac{2\pi}{3} = -\cos\frac{\pi}{3} = -\frac{1}{2}$$

$$\tan\frac{2\pi}{3} = -\tan\frac{\pi}{3} = -\sqrt{3}$$

45. Use a calculator to evaluate (a) $\sin 10°$ and (b) $\csc 10°$ to four decimal places. (Be sure the calculator is set in the correct mode.)

Solution:

(a) $\sin 10° \approx 0.1736$

(b) $\csc 10° = \dfrac{1}{\sin 10°} \approx 5.7588$

49. Use a calculator to evaluate (a) $\cos(-110°)$ and (b) $\cos 250°$ to four decimal places. (Be sure the calculator is set in the correct mode.)

Solution:

(a) $\cos(-110°) = \cos 110° \approx -0.3420$

(b) $\cos 250° \approx -0.3420$

53. Find two values of θ that satisfy (a) $\sin\theta = \frac{1}{2}$ and (b) $\sin\theta = -\frac{1}{2}$. List your answers in degrees $(0° \leq \theta < 360°)$ and radians $(0 \leq \theta < 2\pi)$. Do not use a calculator.

Solution:

(a) $\sin\theta = \dfrac{1}{2} > 0$

θ is in either Quadrant I or Quadrant II.

$\theta = 30° = \dfrac{\pi}{6}$ or $\theta = 150° = \dfrac{5\pi}{6}$

(b) $\sin\theta = -\dfrac{1}{2} < 0$

θ is in either Quadrant III or Quadrant IV.

$\theta = 210° = \dfrac{7\pi}{6}$ or $\theta = 330° = \dfrac{11\pi}{6}$

57. Find two values of θ that satisfy (a) $\tan\theta = 1$ and (b) $\cot\theta = -\sqrt{3}$. List your answers in degrees $(0° \leq \theta < 360°)$ and radians $(0 \leq \theta < 2\pi)$. Do not use a calculator.

Solution:

(a) $\tan\theta = 1$, θ lies in either Quadrant I or Quadrant III.

$\theta = 45° = \dfrac{\pi}{4}$ or $\theta = 225° = \dfrac{5\pi}{4}$

(b) $\cot\theta = -\sqrt{3}$, θ lies in either Quadrant II or Quadrant IV.

$\theta = 150° = \dfrac{5\pi}{6}$ or $\theta = 330° = \dfrac{11\pi}{6}$

65. Evaluate $\sin^2 2 + \cos^2 2$ without using a calculator.

Solution:

Since $\sin^2 \theta + \cos^2 \theta = 1$, we have $\sin^2 2 + \cos^2 2 = 1$.

69. The average daily temperature (in degrees Fahrenheit) for a certain city is given by

$$T = 45 - 23 \cos \left[\frac{2\pi}{365} (t - 32) \right]$$

where t is the time in days with $t = 1$ corresponding to January 1. Find the average temperature on (a) January 1, (b) July 4 ($t = 185$), and (c) October 18 ($t = 291$).

Solution:

(a) $t = 1$

$$T = 45 - 23 \cos \left[\frac{2\pi}{365} (1 - 32) \right] \approx 25.2°F$$

(b) $t = 185$

$$T = 45 - 23 \cos \left[\frac{2\pi}{365} (185 - 32) \right] \approx 65.1°F$$

(c) $t = 291$

$$T = 45 - 23 \cos \left[\frac{2\pi}{365} (291 - 32) \right] \approx 50.8°F$$

SECTION 2.5

Graphs of Sine and Cosine

- You should be able to graph $y = a\sin(bx - c)$ and $y = a\cos(bx - c)$.

- Amplitude: $|a|$

- Period: $\dfrac{2\pi}{|b|}$

- Shift: Solve $bx - c = 0$ and $bx - c = 2\pi$.

- Key Increments: $\dfrac{1}{4}$ (period)

Solutions to Selected Exercises

5. Determine the period and amplitude of $y = \frac{1}{2}\sin \pi x$.

Solution:

$y = \dfrac{1}{2}\sin \pi x; \quad a = \dfrac{1}{2}, \ b = \pi, \ c = 0$

Period: $\dfrac{2\pi}{|b|} = \dfrac{2\pi}{\pi} = 2$

Amplitude: $|a| = \left|\dfrac{1}{2}\right| = \dfrac{1}{2}$

9. Determine the period and amplitude of $y = -2\sin 10x$.

Solution:

$y = -2\sin 10x; \quad a = -2, \ b = 10, \ c = 0$

Period: $\dfrac{2\pi}{|b|} = \dfrac{2\pi}{10} = \dfrac{\pi}{5}$

Amplitude: $|a| = |-2| = 2$

11. Determine the period and amplitude of

$$y = \frac{1}{2}\cos\frac{2x}{3}.$$

Solution:

$y = \dfrac{1}{2}\cos\dfrac{2x}{3}; \quad a = \dfrac{1}{2}, \; b = \dfrac{2}{3}, \; c = 0$

Period: $\dfrac{2\pi}{|b|} = \dfrac{2\pi}{2/3} = 3\pi$

Amplitude: $|a| = \left|\dfrac{1}{2}\right| = \dfrac{1}{2}$

15. Describe the relationship between the graphs of $f(x) = \sin x$ and $g(x) = \sin(x - \pi)$.

Solution:

$f(x) = \sin x$ and $g(x) = \sin(x - \pi)$ both have a period of 2π and an amplitude of 1. However, the graph of $g(x) = \sin(x - \pi)$ is the graph of $f(x) = \sin x$ shifted to the right π units.

19. Describe the relationship between the graphs of $f(x) = \cos x$ and $g(x) = \cos 2x$.

Solution:

$f(x) = \cos x$ and $g(x) = \cos 2x$ both have an amplitude of 1. However, $f(x) = \cos x$ has a period of 2π, whereas $g(x) = \cos 2x$ has a period of π.

23. Sketch the graphs of $f(x) = -2\sin x$ and $g(x) = 4\sin x$ on the same coordinate plane. (Include two full periods.)

Solution:

$f(x) = -2\sin x$
Period: 2π
Amplitude: 2

$g(x) = 4\sin x$
Period: 2π
Amplitude: 4

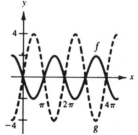

27. Sketch the graphs of the following on the same coordinate plane. (Include two full periods.)

$$f(x) = -\dfrac{1}{2}\sin\dfrac{x}{2} \quad \text{and} \quad g(x) = 3 - \dfrac{1}{2}\sin\dfrac{x}{2}$$

Solution:

$f(x) = -\dfrac{1}{2}\sin\dfrac{x}{2}$

Period: 4π

Amplitude: $\dfrac{1}{2}$

$g(x) = 3 - \dfrac{1}{2}\sin\dfrac{x}{2}$ is the graph of
$f(x)$ shifted vertically three units upward.

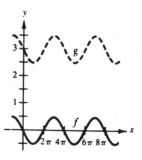

31. Sketch the graph of $y = -2\sin 6x$. (Include two full periods.)

Solution:

$y = -2\sin 6x; \quad a = -2, \; b = 6, \; c = 0$

Period: $\dfrac{2\pi}{6} = \dfrac{\pi}{3}$

Amplitude: $|-2| = 2$

Key points: $(0, \, 0)$, $\left(\dfrac{\pi}{12}, \, -2\right)$, $\left(\dfrac{\pi}{6}, \, 0\right)$, $\left(\dfrac{\pi}{4}, \, 2\right)$, $\left(\dfrac{\pi}{3}, \, 0\right)$

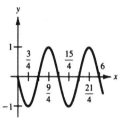

35. Sketch the graph of the following. (Include two full periods.)

$$y = -\sin\frac{2\pi x}{3}$$

Solution:

$y = -\sin\dfrac{2\pi x}{3}; \quad a = -1, \; b = \dfrac{2\pi}{3}, \; c = 0$

Period: $\dfrac{2\pi}{2\pi/3} = 3$

Amplitude: 1

Key points: $(0, \, 0)$, $\left(\dfrac{3}{4}, \, -1\right)$, $\left(\dfrac{3}{2}, \, 0\right)$, $\left(\dfrac{9}{4}, \, 1\right)$, $(3, \, 0)$

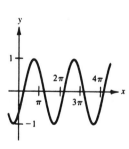

39. Sketch the graph of the following. (Include two full periods.)

$$y = \sin\left(x - \frac{\pi}{4}\right)$$

Solution:

$y = \sin\left(x - \dfrac{\pi}{4}\right); \quad a = 1, \; b = 1, \; c = \dfrac{\pi}{4}$

Period: 2π

Amplitude: 1

Shift: Set $x - \dfrac{\pi}{4} = 0 \quad$ and $\quad x - \dfrac{\pi}{4} = 2\pi$

$\qquad\qquad x = \dfrac{\pi}{4} \qquad\qquad\qquad x = \dfrac{9\pi}{4}$

Key points: $\left(\dfrac{\pi}{4}, \, 0\right)$, $\left(\dfrac{3\pi}{4}, \, 1\right)$, $\left(\dfrac{5\pi}{4}, \, 0\right)$, $\left(\dfrac{7\pi}{4}, \, -1\right)$, $\left(\dfrac{9\pi}{4}, \, 0\right)$

45. Sketch the graph of the following. (Include two full periods.)

$$y = \frac{2}{3}\cos\left(\frac{x}{2} - \frac{\pi}{4}\right)$$

Solution:

$y = \frac{2}{3}\cos\left(\frac{x}{2} - \frac{\pi}{4}\right); \quad a = \frac{2}{3}, \; b = \frac{1}{2}, \; c = \frac{\pi}{4}$

Period: 4π

Amplitude: $\frac{2}{3}$

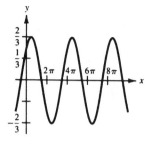

Shift: Set $\dfrac{x}{2} - \dfrac{\pi}{4} = 0$ and $\dfrac{x}{2} - \dfrac{\pi}{4} = 2\pi$

$$x = \frac{\pi}{2} \qquad\qquad x = \frac{9\pi}{2}$$

Key points: $\left(\dfrac{\pi}{2}, \dfrac{2}{3}\right)$, $\left(\dfrac{3\pi}{2}, 0\right)$, $\left(\dfrac{5\pi}{2}, \dfrac{-2}{3}\right)$, $\left(\dfrac{7\pi}{2}, 0\right)$, $\left(\dfrac{9\pi}{2}, \dfrac{2}{3}\right)$

49. Sketch the graph of the following. (Include two full periods.)

$$y = \cos\left(2\pi x - \frac{\pi}{2}\right) + 1$$

Solution:

$y = \cos\left(2\pi x - \dfrac{\pi}{2}\right) + 1; \quad a = 1, \; b = 2\pi, \; c = \dfrac{\pi}{2}$

Period: 1

Amplitude: 1

Shift: Set $2\pi x - \dfrac{\pi}{2} = 0$ and $2\pi x - \dfrac{\pi}{2} = 2\pi$

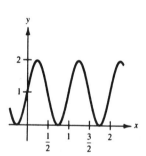

$$x = \frac{1}{4} \qquad\qquad x = \frac{5}{4}$$

Key points: $\left(\dfrac{1}{4}, 2\right)$, $\left(\dfrac{1}{2}, 1\right)$, $\left(\dfrac{3}{4}, 0\right)$, $(1, 1)$, $\left(\dfrac{5}{4}, 2\right)$

Vertical shift: One unit upward

51. Sketch the graph of the following. (Include two full periods.)

$$y = -0.1\sin\left(\frac{\pi x}{10} + \pi\right)$$

Solution:

$y = -0.1 \sin\left(\dfrac{\pi x}{10} + \pi\right); \quad a = -0.1, \ b = \dfrac{\pi}{10}, \ c = -\pi$

Period: $\dfrac{2\pi}{\pi/10} = 20$

Amplitude: $|-0.1| = 0.1$

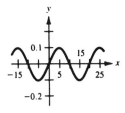

Shift: Set $\dfrac{\pi x}{10} + \pi = 0 \qquad$ and $\qquad \dfrac{\pi x}{10} + \pi = 2\pi$

$\qquad\qquad\qquad\quad x = -10 \qquad\qquad\qquad\qquad x = 10$

Key points: $(-10, \ 0), \ (-5, \ -0.1), \ (0, \ 0), \ (5, \ 0.1), \ (10, \ 0)$

55. Sketch the graph of $y = \frac{1}{10} \cos 60\pi x$. (Include two full periods.)

Solution:

$y = \dfrac{1}{10} \cos(60\pi x); \quad a = \dfrac{1}{10}, \ b = 60\pi, \ c = 0$

Period: $\dfrac{2\pi}{60\pi} = \dfrac{1}{30}$

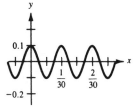

Amplitude: $\dfrac{1}{10}$

Key points: $\left(0, \ \dfrac{1}{10}\right), \ \left(\dfrac{1}{120}, \ 0\right), \ \left(\dfrac{1}{60}, \ -\dfrac{1}{10}\right), \ \left(\dfrac{1}{40}, \ 0\right), \ \left(\dfrac{1}{30}, \ \dfrac{1}{10}\right)$

59. Find a, b, and c so that the graph of the function matches the graph in the figure.

Solution:

$y = a \cos(bx - c)$

Amplitude: $1 \Rightarrow a = 1$

Period: $\dfrac{2\pi}{b} = \pi \Rightarrow b = 2$

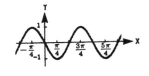

Shift: The graph begins at $-\dfrac{\pi}{4}$.

$2\left(-\dfrac{\pi}{4}\right) - c = 0 \Rightarrow c = -\dfrac{\pi}{2}$

Thus, $y = \cos\left(2x + \dfrac{\pi}{2}\right)$.

65. For a person at rest, the velocity v (in liters per second) of air flow during a respiratory cycle is

$$v = 0.85 \sin \dfrac{\pi t}{3}$$

where t is the time in seconds. (Inhalation occurs when $v > 0$, and exhalation occurs when $v < 0$.)

(a) Find the time for one full respiratory cycle.
(b) Find the number of cycles per minute.
(c) Sketch the graph of the velocity function.

Solution:

(a) Time for one cycle = period = $\dfrac{2\pi}{\pi/3}$ = 6 sec   (b) Cycles per min = $\dfrac{60}{6}$ = 10 cycles per min

(c) Amplitude: 0.85
Period: 6
Key points: $(0,\ 0)$, $\left(\dfrac{3}{2},\ 0.85\right)$, $(3,\ 0)$, $\left(\dfrac{9}{2},\ -0.85\right)$, $(6,\ 0)$

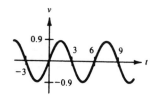

67. When tuning a piano, a technician strikes a tuning fork for the A above middle C and sets up wave motion that can be approximated by

$$y = 0.001 \sin 880\pi t$$

where t is the time in seconds.

(a) What is the period p of this function?
(b) The frequency f is given by $f = 1/p$. What is the frequency of this note?
(c) Sketch the graph of this function.

Solution:

(a) Period: $\dfrac{2\pi}{880\pi} = \dfrac{1}{440}$ (b) $f = \dfrac{1}{p} = 440$

(c) Amplitude: 0.001
Period: $\dfrac{1}{440}$
Key points: $(0,\ 0)$, $\left(\dfrac{1}{1760},\ 0.001\right)$, $\left(\dfrac{1}{880},\ 0\right)$, $\left(\dfrac{3}{1760},\ -0.001\right)$, $\left(\dfrac{1}{440},\ 0\right)$

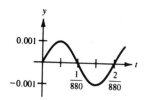

SECTION 2.6

Graphs of Other Trigonometric Functions

■ You should be able to graph

$$y = a \tan(bx - c) \qquad y = a \cot(bx - c)$$
$$y = a \sec(bx - c) \qquad y = a \csc(bx - c)$$

■ When graphing

$$y = a \sec(bx - c) \quad \text{or} \quad y = a \csc(bx - c)$$

you should know to first graph

$$y = a \cos(bx - c) \quad \text{or} \quad y = a \sin(bx - c)$$

since

(a) The intercepts of sine and cosine are vertical asymptotes of cosecant and secant.

(b) The maximum points of sine and cosine are local minimum points of cosecant and secant.

(c) The minimum points of sine and cosine are local maximum points of cosecant and secant.

Solutions to Selected Exercises

5. Match $y = \cot \pi x$ with the correct graph and give the period of the function.

   Solution:
   Period: $\dfrac{\pi}{\pi} = 1$
   Matches graph (d)

   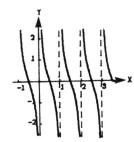

9. Sketch the graph of $y = \tan 2x$ through two periods.

   Solution:
   Period: $\dfrac{\pi}{2}$
   One cycle: $-\dfrac{\pi}{4}$ to $\dfrac{\pi}{4}$

   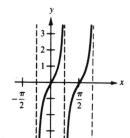

13. Sketch the graph of $y = -2\sec 4x$ through two periods.

Solution:

Period: $\dfrac{2\pi}{4} = \dfrac{\pi}{2}$

One cycle: 0 to $\dfrac{\pi}{2}$

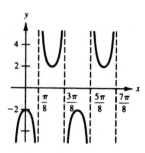

19. Sketch the graph of the following through two periods.

$$y = \csc \frac{x}{2}$$

Solution:

Period: $\dfrac{2\pi}{1/2} = 4\pi$

One cycle 0 to 4π

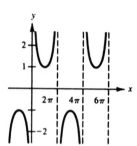

21. Sketch the graph of the following through two periods.

$$y = \cot \frac{x}{2}$$

Solution:

Period: $\dfrac{\pi}{1/2} = 2\pi$

One cycle: 0 to 2π

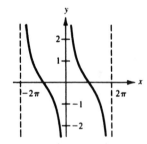

25. Sketch the graph of the following through two periods.

$$y = \tan\left(x - \frac{\pi}{4}\right)$$

Solution:

Period: π

Shift: Set $x - \dfrac{\pi}{4} = -\dfrac{\pi}{2}$ and $x - \dfrac{\pi}{4} = \dfrac{\pi}{2}$

$x = -\dfrac{\pi}{4}$ to $x = \dfrac{3\pi}{4}$

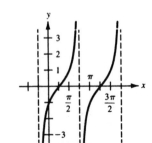

29. Sketch the graph of the following through two periods.

$$y = \frac{1}{4}\cot\left(x - \frac{\pi}{2}\right)$$

Solution:

Period: π

Shift: Set $x - \dfrac{\pi}{2} = 0$ and $x - \dfrac{\pi}{2} = \pi$

$\qquad x = \dfrac{\pi}{2}$ to $\qquad x = \dfrac{3\pi}{2}$

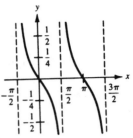

31. Sketch the graph of $y = 2\sec(2x - \pi)$ through two periods.

Solution:

$y = 2\sec(2x - \pi)$

Period: π

Shift: Set $2x - \pi = 0$ and $2x - \pi = 2\pi$

$\qquad x = \dfrac{\pi}{2}$ to $\qquad x = \dfrac{3\pi}{2}$

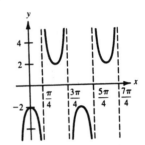

35. Sketch the graph of $y = \csc(\pi - x)$ through two periods.

Solution:

$y = \csc(\pi - x)$

Period: 2π

Shift: Set $\pi - x = 0$ and $\pi - x = 2\pi$

$\qquad x = \pi$ to $\qquad x = -\pi$

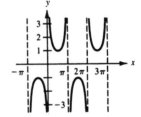

37. A plane flying at an altitude of 6 miles over level ground will pass directly over a radar antenna, as shown in the figure. Let d be the ground distance from the antenna to the point directly under the plane and let x be the angle of elevation to the plane from the antenna. Write d as a function of x, $0 < x < \pi/2$.

Solution:

$$\tan x = \frac{6}{d}$$

$$d = \frac{6}{\tan x} = 6\cot x$$

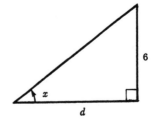

SECTION 2.7

Additional Graphing Techniques

■ You should be able to graph by addition of ordinates.

■ You should be able to graph vertical translations.

■ You should be able to graph using a damping factor.

Solutions to Selected Exercises

1. Use addition of ordinates to sketch the graph of

$$y = 2 - 2\sin\frac{x}{2}.$$

Solution:
Vertical translation of the graph of
$y = -2\sin\dfrac{x}{2}$ by two units

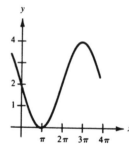

7. Use addition of ordinates to sketch the graph of $y = 1 + \csc x$.

Solution:
Vertical translation of the graph of
$y = 1 + \csc x$ by one unit

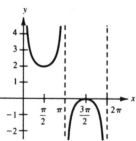

11. Use addition of ordinates to sketch the graph of $y = \frac{1}{2}x - 2\cos x$.

Solution:

$$y = \frac{1}{2}x - 2\cos x$$

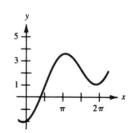

15. Use addition of ordinates to sketch the graph of $y = 2\sin x + \sin 2x$.

Solution:

$$y = 2\sin x + \sin 2x$$

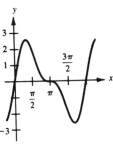

21. Use addition of ordinates to sketch the graph of $y = -3 + \cos x + 2\sin 2x$.

Solution:

$$y = -3 + \cos x + 2\sin 2x$$

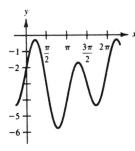

27. Sketch the graph of $y = x\cos x$.

Solution:

$$y = x\cos x$$
$$|x\cos x| = |x||\cos x| \le |x|$$

Thus, the graph lies between the lines $y = -x$ and $y = x$.

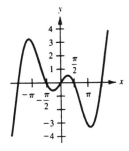

31. Sketch the graph of $y = e^{-x^2/2}\sin x$.

Solution:

$$y = e^{-x^2/2}\sin x$$
$$|e^{-x^2/2}\sin x| = |e^{-x^2/2}||\sin x| \le |e^{-x^2/2}|$$

Thus, $-e^{-x^2/2} \le y \le e^{-x^2/2}$.

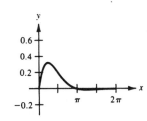

35. The monthly sales S (in thousands of units) of a seasonal product is approximated by $S = 74.50 + 43.75\sin(\pi t/6)$ where t is the time in months with $t = 1$ corresponding to January. Sketch the graph of this sales function over one year.

Solution:

$$S = 74.50 + 43.75 \sin \frac{\pi t}{6}$$

Vertical translation of the graph of
$y = 43.75 \sin \dfrac{\pi t}{6}$ by 74.50 units.

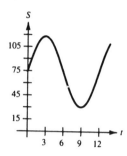

39. Use a calculator to evaluate the function

$$f(x) = \frac{1 - \cos x}{x}$$

at several points in the interval $[-1, \ 1]$, and then use these points to sketch the graph of f. This function is undefined when $x = 0$. From your graph, estimate the value that $f(x)$ is approaching as x approaches 0.

Solution:
Make sure your calculator is in radian mode.

x	−0.5	−0.4	−0.3	−0.2	−0.1	0.1	0.2	0.3	0.4	0.5
$\dfrac{1 - \cos x}{x}$	−0.245	−0.197	−0.149	−0.100	−0.050	0.050	0.100	0.149	0.197	0.245

As $x \to 0$, $f(x) = \dfrac{1 - \cos x}{x} \to 0.$

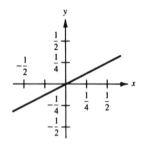

SECTION 2.8

Inverse Trigonometric Functions

- You should know the definitions, domains, and ranges of $y = \arcsin x$, $y = \arccos x$, and $y = \arctan x$.

- You should know the inverse properties of the inverse trigonometric functions.

- You should be able to use the triangle technique to convert trigonometric expressions into algebraic expressions.

Solutions to Selected Exercises

3. Evaluate $\arccos \frac{1}{2}$ without using a calculator.

Solution:

$$\arccos \frac{1}{2} = \theta$$

$$\cos \theta = \frac{1}{2}$$

$$\theta = \frac{\pi}{3}$$

7. Evaluate $\arccos\left(-\frac{\sqrt{3}}{2}\right)$ without using a calculator.

Solution:

$$\arccos\left(-\frac{\sqrt{3}}{2}\right) = \theta$$

$$\cos \theta = -\frac{\sqrt{3}}{2}, \quad \frac{\pi}{2} < \theta < \pi$$

$$\theta = \frac{5\pi}{6}$$

9. Evaluate $\arctan(-\sqrt{3})$ without using a calculator.

Solution:

$$\arctan\left(-\sqrt{3}\right) = \theta$$

$$\tan\theta = -\sqrt{3}, \quad -\frac{\pi}{2} < \theta < 0$$

$$\theta = -\frac{\pi}{3}$$

13. Evaluate $\arcsin \frac{\sqrt{3}}{2}$ without using a calculator.

Solution:

$$\arcsin \frac{\sqrt{3}}{2} = \theta$$

$$\sin\theta = \frac{\sqrt{3}}{2}$$

$$\theta = \frac{\pi}{3}$$

17. Use a calculator to approximate $\arccos 0.28$. (Round your answer to two decimal places.)

Solution:

Make sure that your calculator is in radian mode.

$$\arccos 0.28 \approx 1.29$$

21. Use a calculator to approximate $\arctan(-2)$. (Round your answer to two decimal places.)

Solution:

$$\arctan(-2) \approx -1.11$$

27. Use a calculator to approximate $\arctan 0.92$. (Round your answer to two decimal places.)

Solution:

$$\arctan 0.92 \approx 0.74$$

31. Use the properties of inverse trigonometric functions to evaluate $\cos[\arccos(-0.1)]$.

Solution:

$$\cos[\arccos(-0.1)] = -0.1$$

35. Find the exact value of $\sin\left(\arctan \frac{3}{4}\right)$ without using a calculator. [*Hint:* Make a sketch of a right triangle, as illustrated in Example 6.]

Solution:

Let $y = \arctan \dfrac{3}{4}$. Then,

$$\tan y = \frac{3}{4}, \quad 0 < y < \frac{\pi}{2}$$

and $\sin y = \dfrac{3}{5}$.

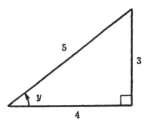

41. Find the exact value of $\sec[\arctan(-\frac{3}{5})]$ without using a calculator. [*Hint:* Make a sketch of a right triangle, as illustrated in Example 6.]

Solution:

Let $y = \arctan\left(-\dfrac{3}{5}\right)$. Then,

$$\tan y = -\frac{3}{5}, \quad -\frac{\pi}{2} < y < 0$$

and $\sec y = \dfrac{\sqrt{34}}{5}$.

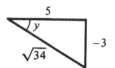

45. Write an algebraic expression that is equivalent to $\cos(\arcsin 2x)$. [*Hint:* Sketch a right triangle, as demonstrated in Example 7.]

Solution:

Let $y = \arcsin(2x)$.

Then, $\sin y = 2x = \dfrac{2x}{1}$

and $\cos y = \sqrt{1 - 4x^2}$.

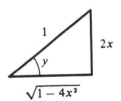

49. Write an algebraic expression that is equivalent to $\tan[\arccos(x/3)]$. [*Hint:* Sketch a right triangle, as demonstrated in Example 7.]

Solution:

Let $y = \arccos \dfrac{x}{3}$.

Then, $\cos y = \dfrac{x}{3}$

and $\tan y = \dfrac{\sqrt{9 - x^2}}{x}$.

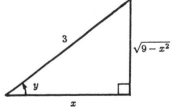

53. Fill in the blank.

$$\arctan \frac{9}{x} = \arcsin(\underline{\hspace{2cm}})$$

Solution:

$$\arctan \frac{9}{x} = \arcsin \frac{9}{\sqrt{x^2 + 81}}$$

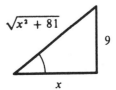

57. Sketch the graph of $f(x) = \arcsin(x - 1)$.

Solution:

The graph of $f(x) = \arcsin(x - 1)$ is a horizontal translation of the graph of $y = \arcsin x$ by one unit.

61. A photographer is taking a picture of a four-foot painting hung in an art gallery. The camera lens is one foot below the lower edge of the painting, as shown in the figure. The angle β subtended by the camera lens x feet from the painting is given by

$$\beta = \arctan \frac{4x}{x^2 + 5}.$$

Find β when (a) $x = 3$ feet and (b) $x = 6$ feet.

Solution:

(a) When $x = 3$,

$$\beta = \arctan\left(\frac{12}{9 + 5}\right)$$

$$\approx 0.7086 \text{ radians or } 40.6°$$

(b) When $x = 6$,

$$\beta = \arctan\left(\frac{24}{36 + 5}\right)$$

$$\approx 0.5296 \text{ radians or } 30.3°$$

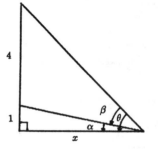

67. Prove the identity of $\arcsin(-x) = -\arcsin x$.

Solution:

Let $y = \arcsin(-x)$. Then,

$$\sin y = -x$$

$$-\sin y = x$$

$$\sin(-y) = x$$

$$-y = \arcsin x$$

$$y = -\arcsin x.$$

Therefore, $\arcsin(-x) = -\arcsin x$.

SECTION 2.9

Applications of Trigonometry

- ■ You should be able to solve right triangles.

- ■ You should be able to solve right triangle applications.

Solutions to Selected Exercises

5. Solve the right triangle, given $A = 12°\,15'$ and $c = 430.5$. (Round your answers to two decimal places.)

Solution:

$A = 12°\,15', \quad c = 430.5$

$B = 90° - 12°\,15' = 77°\,45'$

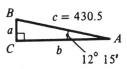

$$\sin 12°\,15' = \frac{a}{430.5}$$

$$a = 430.5 \sin 12°\,15' \approx 91.34$$

$$\cos 12°\,15' = \frac{b}{430.5}$$

$$b = 430.5 \cos 12°\,15' \approx 420.70$$

9. Solve the right triangle, given $b = 16$ and $c = 52$. (Round your answers to two decimal places.)

Solution:

$b = 16, \quad c = 52$

$$a = \sqrt{52^2 - 16^2} = \sqrt{2448} = 12\sqrt{17}$$

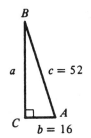

$$\cos A = \frac{16}{52}$$

$$A = \arccos \frac{16}{52} \approx 72.08°$$

$$B = 90 - 72.08 \approx 17.92°$$

13. An isosceles triangle has two angles of 52°, as shown in the figure. The base of the triangle is 4 inches. Find the altitude of the triangle.

Solution:

Divide the triangle in half. Then

$$\tan 52° = \frac{h}{2}$$
$$h = 2 \tan 52°$$
$$\approx 2.56 \text{ inches}$$

15. An amateur radio operator erects a 75-foot vertical tower for his antenna. Find the angle of elevation to the top of the tower at a point on level ground 50 feet from the base.

Solution:

$$\tan \theta = \frac{75}{50}$$
$$\theta = \arctan \frac{3}{2} = 56.3°$$

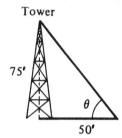

Tower

75'

50'

19. From a point 50 feet in front of a church, the angles of elevation to the base of the steeple and the top of the steeple are 35° and 47°40′, respectively, as shown in the figure. Find the height of the steeple.

Solution:

Let   Height of the church $= x$.
        Height of the church and steeple $= y$.

Then, $\tan 35° = \dfrac{x}{50}$ and $\tan 47°40′ = \dfrac{y}{50}$
           $x = 35.01$ $y = 54.88$

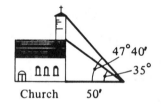

47°40′

35°

Church   50'

Height of the steeple $= y - x = 19.9$ feet

23. A ship is 45 miles east and 30 miles south of port. If the captain wants to travel directly to port, what bearing should be taken?

Solution:

$$\tan A = \frac{30}{45}$$
$$A = \arctan \frac{2}{3}$$
$$A = 30.69°$$

Bearing $= 90 - 30.69 = 56.3° = $ N 56.3° W

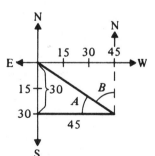

27. An observer in a lighthouse 300 feet above sea level spots two ships directly offshore. The angles of depression to the ships are 4° and 6.5°, as shown in the figure. How far apart are the ships?

Solution:

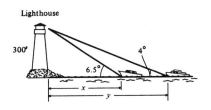

Lighthouse

$$\tan 4° = \frac{300}{y}$$

$$y = \frac{300}{\tan 4°} \approx 4290.20$$

$$\tan 6.5° = \frac{x}{300}$$

$$x = \frac{300}{\tan 6.5} \approx 2633.07$$

Distance between ships = 1657.13 feet

33. Use the figure to find the distance y across the flat sides of the hexagonal nut as a function of r.

Solution:

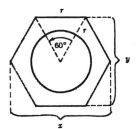

$$\sin 60° = \frac{\text{opp.}}{\text{hyp.}} = \frac{\frac{1}{2}y}{r}$$

$$\frac{\sqrt{3}}{2} = \frac{y}{2r}$$

$$y = \sqrt{3}r$$

37. For the simple harmonic motion described by $d = 4\cos 8\pi t$, find (a) the maximum displacement, (b) the frequency, and (c) the least positive value of t for which $d = 0$.

Solution:
(a) Maximum displacement = amplitude = 4

(b) Frequency $= \dfrac{\omega}{\text{period}} = \dfrac{8\pi}{2\pi} = 4$ cycles per unit of time

(c) $d = 0$ when $8\pi t = \dfrac{\pi}{2}$ or $t = \dfrac{1}{16}$.

41. A point on the end of a tuning fork moves in simple harmonic motion described by $d = a\sin\omega t$. Find ω given that the tuning fork for middle C has a frequency of 264 vibrations per second.

Solution:

$$\text{Frequency} = \frac{\omega}{2\pi} = 264$$

$$\omega = 528\pi$$

REVIEW EXERCISES FOR CHAPTER 2

Solutions to Selected Exercises

3. Sketch the angle $-110°$ in standard position, and list one positive and one negative coterminal angle.

Solution:

$\theta = -110°$

Coterminal angles:

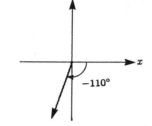

$$\theta_1 = 360° + (-110°) = 250°$$
$$\theta_2 = -360° + (-110°) = -470°$$

7. Convert $5°22'53''$ to decimal form. Round your answer to two decimal places.

Solution:

$$5°22'53'' = 5 + \frac{22}{60} + \frac{53}{3600} \approx 5.38°$$

11. Convert $-85.15°$ to $D° \; M' \; S''$ form.

Solution:

$$-85.15° = -[85° + .15(60)'] = -85°9'$$

15. Convert -3.5 from radians to degrees. Round your answer to two decimal places.

Solution:

$$-3.5 = -3.5\left(\frac{180}{\pi}\right)° \approx -200.54°$$

19. Convert $-33°45'$ from degrees to radians. Round your answer to four decimal places.

Solution:

$$-33°45' = -\left[33 + \frac{45}{60}\right]°$$
$$= -33.75°$$
$$= -33.75\left(\frac{\pi}{180}\right)$$
$$\approx -0.5890 \text{ radians}$$

23. Find the reference angle for 252°.

Solution:

252° is in Quadrant III. The reference angle is

$$\theta = 252° - 180° = 72°.$$

27. Find the six trigonometric functions of the angle θ (in standard position) whose terminal side passes through the point $(-4, -6)$.

Solution:

$x = -4, \; y = -6, \; r = \sqrt{(-4)^2 + (-6)^2} = 2\sqrt{13}$

$$\sin\theta = \frac{y}{r} = \frac{-6}{2\sqrt{13}} = -\frac{3\sqrt{13}}{13} \qquad \csc\theta = \frac{r}{y} = \frac{2\sqrt{13}}{-6} = -\frac{\sqrt{13}}{3}$$

$$\cos\theta = \frac{x}{r} = \frac{-4}{2\sqrt{13}} = -\frac{2\sqrt{13}}{13} \qquad \sec\theta = \frac{r}{x} = \frac{2\sqrt{13}}{-4} = -\frac{\sqrt{13}}{2}$$

$$\tan\theta = \frac{y}{x} = \frac{-6}{-4} = \frac{3}{2} \qquad \cot\theta = \frac{x}{y} = \frac{-4}{-6} = \frac{2}{3}$$

31. Use a right triangle to find the remaining five trigonometric functions of θ, given $\sin\theta = \frac{3}{8}$ and $\cos\theta < 0$.

Solution:

$\sin\theta = \frac{3}{8}, \quad \cos\theta < 0, \; \theta$ is in Quadrant II

$y = 3, \; r = 8, \; x = -\sqrt{64 - 9} = -\sqrt{55}$

$$\cos\theta = -\frac{\sqrt{55}}{8} \qquad \sec\theta = \frac{8}{-\sqrt{55}} = -\frac{8\sqrt{55}}{55}$$

$$\tan\theta = \frac{3}{-\sqrt{55}} = -\frac{3\sqrt{55}}{55} \qquad \cot\theta = -\frac{\sqrt{55}}{3}$$

$$\csc\theta = \frac{8}{3}$$

35. Evaluate $\cos 495°$ without the use of a calculator.

Solution:

The reference angle for 495° is 45° and 495° is in Quadrant II. Therefore,

$$\cos 495° = -\cos 45° = -\frac{\sqrt{2}}{2}.$$

39. Use a calculator to evaluate $\sec(12\pi/5)$. Round your answer to two decimal places.

Solution:

Make sure that your calculator is in radian mode.

$$\sec\left(\frac{12\pi}{5}\right) = \frac{1}{\cos(12\pi/5)} \approx 3.24$$

43. Given $\csc \theta = -2$, find two values of θ in degrees $(0° \leq \theta < 360°)$ and in radians $(0 \leq \theta < 2\pi)$ without using a calculator.

Solution:

Since $\csc \theta < 0$, we know that θ is in either Quadrant III or in Quadrant IV. Also, since $\csc 30° = 2$, we know that $30°$ is the reference angle.

In Quadrant III: $\theta = 180° + 30° = 210° = \dfrac{7\pi}{6}$

In Quadrant IV: $\theta = 360° - 30° = 330° = \dfrac{11\pi}{6}$

47. Given $\sec \theta = -1.0353$, find two values of θ in degrees $(0° \leq \theta < 360°)$ and in radians $(0 \leq \theta < 2\pi)$ by using a calculator.

Solution:

Since $\sec \theta < 0$, we know that θ is in either Quadrant II or in Quadrant III. To find the reference angle θ', use

$$\cos \theta' = \frac{1}{\sec \theta'} = \frac{1}{1.0353}$$

$$\theta' = \arccos\left(\frac{1}{1.0353}\right)$$

The reference angle θ' is $15°$.

In Quadrant II: $\theta = 180° - 15° = 165° \approx 2.8798$

In Quadrant III: $\theta = 180° + 15° = 195° \approx 3.4034$

51. Use a right triangle to write an algebraic expression for

$$\sin\left(\arccos \frac{x^2}{4 - x^2}\right).$$

Solution:

Find $\sin\left(\arccos \dfrac{x^2}{4 - x^2}\right)$.

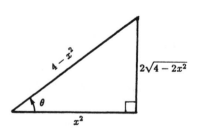

Let $\theta = \dfrac{x^2}{4 - x^2}$, then, $\cos \theta = \dfrac{x^2}{4 - x^2}$.

The opposite side is $\sqrt{(4 - x^2)^2 - (x^2)^2} = \sqrt{16 - 8x^2} = 2\sqrt{4 - 2x^2}$.

Thus, $\sin \theta = \dfrac{2\sqrt{4 - 2x^2}}{4 - x^2}$.

55. Sketch the graph of

$$f(x) = -\frac{1}{4} \cos \frac{\pi x}{4}.$$

Solution:

Amplitude: $\left| -\frac{1}{4} \right| = \frac{1}{4}$

Period: $\dfrac{2\pi}{\pi/4} = 8$

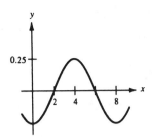

59. Sketch the graph of

$$h(t) = \csc\left(3t - \frac{\pi}{2}\right).$$

Solution:

Period: $\dfrac{2\pi}{3}$

Shift: $3t - \dfrac{\pi}{2} = 0$ and $3t - \dfrac{\pi}{2} = 2\pi$

$t = \dfrac{\pi}{6}$ $\qquad\qquad t = \dfrac{5\pi}{6}$

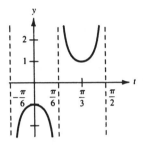

63. Sketch the graph of

$$f(x) = \frac{x}{4} - \sin x.$$

Solution:

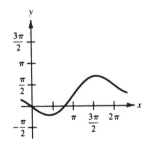

67. Sketch the graph of

$$f(x) = \arcsin \frac{x}{2}.$$

Solution:

Domain: $-2 \le x \le 2$

Range: $-\dfrac{\pi}{2} \le f(x) \le \dfrac{\pi}{2}$

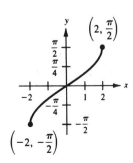

71. An observer 2.5 miles from the launch pad of a space shuttle measures the angle of elevation to the base of the vehicle to be 28° soon after liftoff (see figure). How high is the shuttle at that instant? Assume that the shuttle is still moving vertically.

Solution:

$$\tan 28° = \frac{x}{2.5}$$

$$x = 2.5 \tan 28° \approx 1.33 \text{ miles}$$

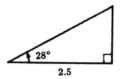

Practice Test for Chapter 2

1. (a) Express 350° in radian measure. (b) Express $\dfrac{5\pi}{9}$ in degree measure.

2. (a) Convert 135°14′12″ to decimal form. (b) Convert $-22.569°$ to D° M′ S″ form.

3. Use the unit circle to evaluate

  (a) $\sin \dfrac{5\pi}{6}$ (b) $\tan \dfrac{5\pi}{4}$

4. Use the unit circle and the periodic nature of sine and cosine to evaluate

  (a) $\sin 7\pi$ (b) $\cos\left(-\dfrac{13\pi}{3}\right)$

5. If $\cos\theta = \frac{2}{3}$, use the trigonometric identities to find $\tan\theta$.

6. Find θ given $\sin\theta = 0.9063$.

7. Solve for x.

8. Find the magnitude of the reference angle for $\theta = \dfrac{6\pi}{5}$.

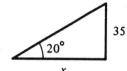

9. Evaluate csc 3.92.

10. Find $\sec\theta$ given that θ lies in Quadrant III and $\tan\theta = 6$.

11. Graph $y = 3\sin\dfrac{x}{2}$. **12.** Graph $y = -2\cos(x - \pi)$.

13. Graph $y = \tan 2x$. **14.** Graph $y = -\csc\left(x + \dfrac{\pi}{4}\right)$.

15. Graph $y = 2x + \sin x$. **16.** Graph $y = 3x\cos x$.

17. Evaluate arcsin 1. **18.** Evaluate $\arctan(-3)$.

19. Evaluate $\sin\left(\arccos\dfrac{4}{\sqrt{35}}\right)$. **20.** Write an algebraic expression for $\cos\left(\arcsin\dfrac{x}{4}\right)$.

For Exercises 21–23, solve the right triangle.

21. $A = 40°$, $c = 12$

22. $B = 6.84°$, $a = 21.3$

23. $a = 5$, $b = 9$

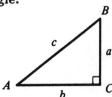

24. A 20-foot ladder leans against the side of a barn. Find the height of the top of the ladder if the angle of elevation of the ladder is 67°.

25. An observer in a lighthouse 250 feet above sea level spots a ship off the shore. If the angle of depression to the ship is 5°, how far out is the ship?

CHAPTER 3

Analytic Trigonometry

SECTION 3.1

Applications of Fundamental Identities

- You should know the following identities.

  (a) Reciprocal Identities
  (b) Tangent and Cotangent Identities
  (c) Pythagorean Identities
  (d) Cofunction Identities
  (e) Negative Angle Identities

- You should be able to use these fundamental identities to find function values.

- You should be able to convert trigonometric expressions to equivalent forms by using the fundamental identities.

Solutions to Selected Exercises

3. Given $\sec \theta = \sqrt{2}$ and $\sin \theta = -\sqrt{2}/2$, use the fundamental identities to find the values of the other four trigonometric functions.

Solution:

$$\cos \theta = \frac{1}{\sec \theta} = \frac{1}{\sqrt{2}} = \frac{\sqrt{2}}{2}$$

$$\tan \theta = \frac{\sin \theta}{\cos \theta} = \frac{-\sqrt{2}/2}{\sqrt{2}/2} = -1$$

$$\cot \theta = \frac{1}{\tan \theta} = \frac{1}{-1} = -1$$

$$\csc \theta = \frac{1}{\sin \theta} = \frac{1}{-\sqrt{2}/2} = -\sqrt{2}$$

7. Given $\sec \phi = -1$ and $\sin \phi = 0$, use the fundamental identities to find the values of the other four trigonometric functions.

Solution:

$$\cos \phi = \frac{1}{\sec \phi} = \frac{1}{-1} = -1$$

$$\tan \phi = \frac{\sin \phi}{\cos \phi} = \frac{0}{-1} = 0$$

$$\cot \phi = \frac{1}{\tan \phi} = \frac{1}{0} \text{ undefined}$$

$$\csc \phi = \frac{1}{\sin \phi} = \frac{1}{0} \text{ undefined}$$

11. Given $\tan \theta = 2$ and $\sin \theta < 0$, use the fundamental identities to find the values of the other four trigonometric functions.

Solution:

θ is in Quadrant III.

$$\sec \theta = -\sqrt{1 + \tan^2 \theta} = -\sqrt{1 + (2)^2} = -\sqrt{5}$$

$$\cot \theta = \frac{1}{\tan \theta} = \frac{1}{2}$$

$$\csc \theta = -\sqrt{1 + \cot^2 \theta} = -\sqrt{1 + (1/2)^2} = -\sqrt{1 + (1/4)} = -\frac{\sqrt{5}}{2}$$

$$\sin \theta = \frac{1}{\csc \theta} = -\frac{2}{\sqrt{5}} = -\frac{2\sqrt{5}}{5}$$

$$\cos \theta = \frac{1}{\sec \theta} = -\frac{1}{\sqrt{5}} = -\frac{\sqrt{5}}{5}$$

17. Match $\tan^2 x - \sec^2 x$ with one of the following.

(a) -1 (b) $\cos x$ (c) $\cot x$

(d) 1 (e) $-\tan x$ (f) $\sin x$

Solution:

$$\tan^2 x - \sec^2 x = (\sec^2 x - 1) - \sec^2 x = -1$$

Matches (a).

21. Match $\sin x \sec x$ with one of the following.

(a) $\csc x$ (b) $\tan x$ (c) $\sin^2 x$

(d) $\sin x \tan x$ (e) $\sec^2 x$ (f) $\sec^2 x + \tan^2 x$

Solution:

$$\sin x \sec x = \sin x \left(\frac{1}{\cos x}\right) = \frac{\sin x}{\cos x} = \tan x$$

Matches (b).

25. Match $\sec^4 x - \tan^4 x$ with one of the following.

(a) $\csc x$ (b) $\tan x$ (c) $\sin^2 x$

(d) $\sin x \tan x$ (e) $\sec^2 x$ (f) $\sec^2 x + \tan^2 x$

Solution:

$$\begin{aligned}
\sec^4 x - \tan^4 x &= \left(\sec^2 x + \tan^2 x\right)\left(\sec^2 x - \tan^2 x\right)\\
&= \left(\sec^2 x + \tan^2 x\right)\left[\left(1 + \tan^2 x\right) - \tan^2 x\right]\\
&= \sec^2 x + \tan^2 x
\end{aligned}$$

Matches (f).

29. Use the fundamental identities to simplify $\cos \beta \tan \beta$.

Solution:

$$\cos \beta \tan \beta = \cos \beta \left(\frac{\sin \beta}{\cos \beta}\right) = \sin \beta$$

33. Use the fundamental identities to simplify $\sec^2 x\left(1 - \sin^2 x\right)$.

Solution:

$$\sec^2 x\left(1 - \sin^2 x\right) = \sec^2 x\left(\cos^2 x\right) = \frac{1}{\cos^2 x}\left(\cos^2 x\right) = 1$$

39. Use the fundamental identities to simplify

$$\frac{\cos^2 y}{1 - \sin y}.$$

Solution:

$$\frac{\cos^2 y}{1 - \sin y} = \frac{1 - \sin^2 y}{1 - \sin y} = \frac{(1 + \sin y)(1 - \sin y)}{1 - \sin y} = 1 + \sin y$$

43. Factor $\sin^2 x \sec^2 x - \sin^2 x$ and use the fundamental identities to simplify the result.

Solution:

$$\sin^2 x \sec^2 x - \sin^2 x = \sin^2 x\left(\sec^2 x - 1\right) = \sin^2 x \tan^2 x$$

47. Factor $\sin^4 x - \cos^4 x$ and use the fundamental identities to simplify the result.

Solution:

$$\sin^4 x - \cos^4 x = (\sin^2 x + \cos^2 x)(\sin^2 x - \cos^2 x) = 1(\sin^2 x - \cos^2 x) = \sin^2 x - \cos^2 x$$

51. Multiply $(\sec x + 1)(\sec x - 1)$ and use the fundamental identities to simplify the result.

Solution:

$$(\sec x + 1)(\sec x - 1) = \sec^2 x - 1 = \tan^2 x$$

55. Add the following and use the fundamental identities to simplify the result.

$$\frac{\cos x}{1 + \sin x} + \frac{1 + \sin x}{\cos x}$$

Solution:

$$\frac{\cos x}{1 + \sin x} + \frac{1 + \sin x}{\cos x} = \frac{(\cos x)(\cos x) + (1 + \sin x)(1 + \sin x)}{\cos x(1 + \sin x)}$$

$$= \frac{\cos^2 x + 1 + 2\sin x + \sin^2 x}{\cos x(1 + \sin x)}$$

$$= \frac{(\sin^2 x + \cos^2 x) + 1 + 2\sin x}{\cos x(1 + \sin x)}$$

$$= \frac{1 + 1 + 2\sin x}{\cos x(1 + \sin x)}$$

$$= \frac{2(1 + \sin x)}{\cos x(1 + \sin x)}$$

$$= \frac{2}{\cos x}$$

$$= 2\left(\frac{1}{\cos x}\right)$$

$$= 2\sec x$$

59. Rewrite the following expression so that it is *not* in fractional form.

$$\frac{3}{\sec x - \tan x}$$

Solution:

$$\frac{3}{\sec x - \tan x} = \frac{3}{\sec x - \tan x} \cdot \frac{\sec x + \tan x}{\sec x + \tan x} = \frac{3(\sec x + \tan x)}{\sec^2 x - \tan^2 x}$$

$$= \frac{3(\sec x + \tan x)}{(1 + \tan^2 x) - \tan^2 x} = 3(\sec x + \tan x)$$

63. Use $x = 3\sec\theta$ to write $\sqrt{x^2 - 9}$ as a trigonometric function involving θ, where $0 < \theta < \pi/2$.

Solution:

Since $x = 3\sec\theta$,

$$\begin{aligned}
\sqrt{x^2 - 9} &= \sqrt{(3\sec\theta)^2 - 9} \\
&= \sqrt{9\sec^2\theta - 9} \\
&= \sqrt{9(\sec^2\theta - 1)} \\
&= \sqrt{9\tan^2\theta} \\
&= 3\tan\theta
\end{aligned}$$

69. Use $x = 3\tan\theta$ to write $\sqrt{(9 + x^2)^3}$ as a trigonometric function involving θ, where $0 < \theta < \pi/2$.

Solution:

Since $x = 3\tan\theta$,

$$\begin{aligned}
\sqrt{(9 + x^2)^3} &= \sqrt{[9 + (3\tan\theta)^2]^3} \\
&= \sqrt{(9 + 9\tan^2\theta)^3} \\
&= \sqrt{[9(1 + \tan^2\theta)]^3} \\
&= \sqrt{(9\sec^2\theta)^3} \\
&= (\sqrt{9\sec^2\theta})^3 \\
&= (3\sec\theta)^3 \\
&= 27\sec^3\theta
\end{aligned}$$

73. Rewrite $\ln|\cos\theta| - \ln|\sin\theta|$ as a single logarithm and simplify.

Solution:

$$\ln|\cos\theta| - \ln|\sin\theta| = \ln\frac{|\cos\theta|}{|\sin\theta|} = \ln|\cot\theta|$$

75. Determine whether the following statement is true or false, and give a reason for your answer.

$$\frac{\sin k\theta}{\cos k\theta} = \tan\theta, \quad k \text{ is constant}$$

Solution:

$\dfrac{\sin k\theta}{\cos k\theta} = \tan\theta$ is false, since $\dfrac{\sin k\theta}{\cos k\theta} = \tan k\theta$.

79. Use a calculator to demonstrate that $\csc^2 \theta - \cot^2 \theta = 1$ is true for

(a) $\theta = 132°$ (b) $\theta = 2\pi/7$.

Solution:

(a) $\theta = 132°$

$\csc 132° \approx 1.3456$

$\cot 132° \approx -0.9004$

Therefore, $\csc^2 132° - \cot^2 132° \approx 1.8107 - 0.8107 = 1$.

(b) $\theta = \dfrac{2\pi}{7}$

$\csc \dfrac{2\pi}{7} \approx 1.2790$

$\cot \dfrac{2\pi}{7} \approx 0.7975$

Therefore, $\csc^2 \dfrac{2\pi}{7} - \cot^2 \dfrac{2\pi}{7} \approx 1.6360 - 0.6360 = 1$.

83. Express each of the other trigonometric functions of θ in terms of $\sin \theta$.

Solution:

Since $\sin^2 \theta + \cos^2 \theta = 1$ and $\cos^2 \theta = 1 - \sin^2 \theta$,

$$\cos \theta = \pm\sqrt{1 - \sin^2 \theta}$$

$$\tan \theta = \frac{\sin \theta}{\cos \theta} = \frac{\pm \sin \theta}{\sqrt{1 - \sin^2 \theta}}$$

$$\cot \theta = \frac{1}{\tan \theta} = \frac{\pm\sqrt{1 - \sin^2 \theta}}{\sin \theta}$$

$$\sec \theta = \frac{1}{\cos \theta} = \frac{\pm 1}{\sqrt{1 - \sin^2 \theta}}$$

$$\csc \theta = \frac{1}{\sin \theta}$$

SECTION 3.2
Verifying Trigonometric Identities

- ■ You should know the difference between an expression, an equation, and an identity.

- ■ You should be able to solve trigonometric identities, using the following techniques.

  (a) Work with *one* side at a time. Do not "cross" the equal sign.
  (b) Use algebraic techniques such as combining fractions, factoring expressions, rationalizing denominators, and squaring binomials.
  (c) Use the fundamental identities.
  (d) Convert all the terms into sines and cosines.

Solutions to Selected Exercises

5. Verify $\cos^2 \beta - \sin^2 \beta = 1 - 2\sin^2 \beta$.

Solution:

$$\cos^2 \beta - \sin^2 \beta = (1 - \sin^2 \beta) - \sin^2 \beta = 1 - 2\sin^2 \beta$$

9. Verify $\sin^2 \alpha - \sin^4 \alpha = \cos^2 \alpha - \cos^4 \alpha$.

Solution:

$$\sin^2 \alpha - \sin^4 \alpha = \sin^2 \alpha(1 - \sin^2 \alpha) = (1 - \cos^2 \alpha)(\cos^2 \alpha) = \cos^2 \alpha - \cos^4 \alpha$$

13. Verify

$$\frac{\cot^2 t}{\csc t} = \csc t - \sin t.$$

Solution:

$$\frac{\cot^2 t}{\csc t} = \frac{\csc^2 t - 1}{\csc t} = \frac{\csc^2 t}{\csc t} - \frac{1}{\csc t} = \csc t - \sin t$$

17. Verify

$$\frac{1}{\sec x \tan x} = \csc x - \sin x.$$

Solution:

$$\frac{1}{\sec x \tan x} = \frac{1}{\sec x} \cdot \frac{1}{\tan x}$$
$$= \cos x \cot x$$
$$= \cos x \left(\frac{\cos x}{\sin x}\right)$$
$$= \frac{\cos^2 x}{\sin x}$$
$$= \frac{1 - \sin^2 x}{\sin x}$$
$$= \frac{1}{\sin x} - \frac{\sin^2 x}{\sin x}$$
$$= \csc x - \sin x$$

21. Verify $\csc x - \sin x = \cos x \cot x$.

Solution:

$$\csc x - \sin x = \frac{1}{\sin x} - \frac{\sin^2 x}{\sin x} = \frac{1 - \sin^2 x}{\sin x} = \frac{\cos^2 x}{\sin x} = \cos x \left(\frac{\cos x}{\sin x}\right) = \cos x \cot x$$

25. Verify

$$\frac{\cos \theta \cot \theta}{1 - \sin \theta} - 1 = \csc \theta.$$

Solution:

$$\frac{\cos \theta \cot \theta}{1 - \sin \theta} - 1 = \frac{\cos \theta \left(\dfrac{\cos \theta}{\sin \theta}\right)}{1 - \sin \theta} \cdot \frac{1 + \sin \theta}{1 + \sin \theta} - 1$$
$$= \frac{\cos^2 \theta (1 + \sin \theta)}{\sin \theta (1 - \sin^2 \theta)} - 1$$
$$= \frac{\cos^2 \theta (1 + \sin \theta)}{\sin \theta \cos^2 \theta} - 1$$
$$= \frac{1 + \sin \theta}{\sin \theta} - 1$$
$$= \frac{1}{\sin \theta} + \frac{\sin \theta}{\sin \theta} - 1$$
$$= \csc \theta + 1 - 1$$
$$= \csc \theta$$

29. Verify $2\sec^2 x - 2\sec^2 x \sin^2 x - \sin^2 x - \cos^2 x = 1$.

Solution:

$$
\begin{aligned}
2\sec^2 x - 2\sec^2 x \sin^2 x - \sin^2 x - \cos^2 x &= 2\sec^2 x(1 - \sin^2 x) - (\sin^2 x + \cos^2 x) \\
&= 2\sec^2 x(\cos^2 x) - 1 \\
&= 2\left(\frac{1}{\cos^2 x}\right)(\cos^2 x) - 1 \\
&= 2 - 1 = 1
\end{aligned}
$$

33. Verify $\csc^4 x - 2\csc^2 x + 1 = \cot^4 x$.

Solution:

$$\csc^4 x - 2\csc^2 x + 1 = (\csc^2 x - 1)^2 = (\cot^2 x)^2 = \cot^4 x$$

37. Verify

$$\frac{\sin \beta}{1 - \cos \beta} = \frac{1 + \cos \beta}{\sin \beta}.$$

Solution:

$$\frac{\sin \beta}{1 - \cos \beta} = \frac{\sin \beta}{1 - \cos \beta} \cdot \frac{1 + \cos \beta}{1 + \cos \beta} = \frac{\sin \beta(1 + \cos \beta)}{1 - \cos^2 \beta} = \frac{\sin \beta(1 + \cos \beta)}{\sin^2 \beta} = \frac{1 + \cos \beta}{\sin \beta}$$

41. Verify

$$\cos\left(\frac{\pi}{2} - x\right)\csc x = 1.$$

Solution:

$$\cos\left(\frac{\pi}{2} - x\right)\csc x = \sin x \csc x = \sin x\left(\frac{1}{\sin x}\right) = 1$$

45. Verify

$$\frac{\cos(-\theta)}{1 + \sin(-\theta)} = \sec \theta + \tan \theta.$$

Solution:

$$\frac{\cos(-\theta)}{1+\sin(-\theta)} = \frac{\cos\theta}{1-\sin\theta}$$

$$= \frac{\cos\theta}{1-\sin\theta} \cdot \frac{1+\sin\theta}{1+\sin\theta}$$

$$= \frac{\cos\theta(1+\sin\theta)}{1-\sin^2\theta}$$

$$= \frac{\cos\theta(1+\sin\theta)}{\cos^2\theta}$$

$$= \frac{1+\sin\theta}{\cos\theta}$$

$$= \frac{1}{\cos\theta} + \frac{\sin\theta}{\cos\theta}$$

$$= \sec\theta + \tan\theta$$

49. Verify

$$\frac{\tan x + \cot y}{\tan x \cot y} = \tan y + \cot x.$$

Solution:

$$\frac{\tan x + \cot y}{\tan x \cot y} = \frac{\tan x}{\tan x \cot y} + \frac{\cot y}{\tan x \cot y} = \frac{1}{\cot y} + \frac{1}{\tan x} = \tan y + \cot x$$

53. Verify $\ln|\tan\theta| = \ln|\sin\theta| - \ln|\cos\theta|$.

Solution:

$$\ln|\tan\theta| = \ln\left|\frac{\sin\theta}{\cos\theta}\right| = \ln\frac{|\sin\theta|}{|\cos\theta|} = \ln|\sin\theta| - \ln|\cos\theta|$$

57. Verify $\sin^2 x + \sin^2\left(\frac{\pi}{2} - x\right) = 1$.

Solution:

$$\sin^2 x + \sin^2\left(\frac{\pi}{2} - x\right) = \sin^2 x + \cos^2 x = 1$$

63. Explain why $\sqrt{\tan^2 x} = \tan x$ is *not* an identity and find one value of the variable for which the equation is not true.

Solution:

$$\sqrt{\tan^2 x} = |\tan x|$$

To show that $\sqrt{\tan^2 x} = \tan x$ is not true, pick any value of x whose tangent is negative. For example, $\sqrt{\tan^2 135°} = 1$, whereas, $\tan 135° = -1$.

SECTION 3.3

Solving Trigonometric Equations

- You should be able to identify and solve trigonometric equations.

- A trigonometric equation is a conditional equation. It is true for a specific set of values.

- To solve trigonometric equations, use algebraic techniques such as collecting like terms, taking square roots, factoring, squaring, converting to quadratic form, and using formulas.

Solutions to Selected Exercises

5. Verify that the given values of x are solutions of the equation $2\sin^2 x - \sin x - 1 = 0$.

(a) $x = \dfrac{\pi}{2}$

(b) $x = \dfrac{7\pi}{6}$

Solution:

(a) $x = \dfrac{\pi}{2}$

$\sin \dfrac{\pi}{2} = 1$

$2\sin^2 \dfrac{\pi}{2} - \sin \dfrac{\pi}{2} - 1 = 2(1)^2 - 1 - 1 = 0$

(b) $x = \dfrac{7\pi}{6}$

$\sin \dfrac{7\pi}{6} = -\dfrac{1}{2}$

$2\sin^2 \dfrac{7\pi}{6} - \sin \dfrac{7\pi}{6} - 1 = 2\left(-\dfrac{1}{2}\right)^2 - \left(-\dfrac{1}{2}\right) - 1$

$= 2\left(\dfrac{1}{4}\right) + \dfrac{1}{2} - 1$

$= \dfrac{1}{2} + \dfrac{1}{2} - 1$

$= 0$

9. Find all solutions of $\sqrt{3}\csc x - 2 = 0$. (Do not use a calculator.)

Solution:

$$\sqrt{3}\csc x - 2 = 0$$

$$\csc x = \frac{2}{\sqrt{3}}$$

$$x = \frac{\pi}{3} \text{ or } \frac{2\pi}{3} \text{ in } [0,\ 2\pi)$$

In general form, $x = \dfrac{\pi}{3} + 2n\pi$ or $x = \dfrac{2\pi}{3} + 2n\pi$ where n is an integer.

13. Find all solutions of $3\sec^2 x - 4 = 0$. (Do not use a calculator.)

Solution:

$$3\sec^2 x - 4 = 0$$

$$\sec^2 x = \frac{4}{3}$$

$$\sec x = \pm\sqrt{\frac{4}{3}} = \pm\frac{2}{\sqrt{3}} = \pm\frac{2\sqrt{3}}{3}$$

$$\sec x = \frac{2\sqrt{3}}{3} \qquad \text{OR} \qquad \sec x = -\frac{2\sqrt{3}}{3}$$

$$x = \frac{\pi}{6},\ \frac{11\pi}{6} \qquad\qquad x = \frac{5\pi}{6},\ \frac{7\pi}{6} \text{ in } [0,\ 2\pi)$$

In general form, the solutions are

$$x = \frac{\pi}{6} + n\pi \text{ or } x = \frac{5\pi}{6} + n\pi$$

where n is an integer.

17. Find all solutions of $\sin x(\sin x + 1) = 0$. (Do not use a calculator.)

Solution:

$$\sin x = 0 \qquad \text{OR} \qquad \sin x = -1$$

$$x = 0,\ \pi \qquad\qquad x = \frac{3\pi}{2} \text{ in } [0,\ 2\pi)$$

In general form, the solutions are

$$x = n\pi \quad \text{and} \quad x = \frac{3\pi}{2} + 2n\pi,$$

where n is an integer.

21. Find all solutions of $\sec x \csc x - 2 \csc x = 0$ in the interval $[0, 2\pi)$. (Do not use a calculator.)

Solution:

$$\sec x \csc x - 2 \csc x = 0$$
$$\csc x(\sec x - 2) = 0$$

$\csc x = 0$ OR $\sec x = 2$

Not possible $x = \dfrac{\pi}{3}, \dfrac{5\pi}{3}$

25. Find all solutions of $\cos^3 x = \cos x$ in the interval $[0, 2\pi)$. (Do not use a calculator.)

Solution:

$$\cos^3 x = \cos x$$
$$\cos^3 x - \cos x = 0$$
$$\cos x(\cos^2 x - 1) = 0$$

$\cos x = 0$ OR $\cos^2 x - 1 = 0$

$x = \dfrac{\pi}{2}, \dfrac{3\pi}{2}$ $\cos x = \pm 1$

$x = 0, \dfrac{\pi}{2}, \pi, \dfrac{3\pi}{2}$ $x = 0, \pi$

29. Find all solutions of $2\sec^2 x + \tan^2 x - 3 = 0$ in the interval $[0, 2\pi)$. (Do not use a calculator.)

Solution:

$$2\sec^2 x + \tan^2 x - 3 = 0$$
$$2\sec^2 x + (\sec^2 x - 1) - 3 = 0$$
$$3\sec^2 x - 4 = 0$$
$$\sec^2 x = \frac{4}{3}$$
$$\sec x = \pm \frac{2}{\sqrt{3}}$$

$\sec x = \dfrac{2}{\sqrt{3}}$ OR $\sec x = -\dfrac{2}{\sqrt{3}}$

$x = \dfrac{\pi}{6}, \dfrac{11\pi}{6}$ $x = \dfrac{5\pi}{6}, \dfrac{7\pi}{6}$

$x = \dfrac{\pi}{6}, \dfrac{5\pi}{6}, \dfrac{7\pi}{6}, \dfrac{11\pi}{6}$

33. Find all solutions of $\sin 2x = -\sqrt{3}/2$ in the interval $[0,\ 2\pi)$. (Do not use a calculator.)

Solution:

$$2x = \frac{4\pi}{3}, \quad 2x = \frac{5\pi}{3}, \quad 2x = \frac{10\pi}{3}, \quad 2x = \frac{11\pi}{3}$$

$$x = \frac{2\pi}{3}, \quad x = \frac{5\pi}{6}, \quad x = \frac{5\pi}{3}, \quad x = \frac{11\pi}{6}$$

39. Find all solutions of

$$\frac{1 + \sin x}{\cos x} + \frac{\cos x}{1 + \sin x} = 4$$

in the interval $[0,\ 2\pi)$. (Do not use a calculator.)

Solution:

$$\frac{1 + \sin x}{\cos x} + \frac{\cos x}{1 + \sin x} = 4$$
$$(1 + \sin x)^2 + (\cos x)^2 = 4\cos x(1 + \sin x)$$
$$1 + 2\sin x + \sin^2 x + \cos^2 x = 4\cos x(1 + \sin x)$$
$$1 + 2\sin x + 1 = 4\cos x(1 + \sin x)$$
$$2(1 + \sin x) - 4\cos(1 + \sin x) = 0$$
$$2(1 + \sin x)(1 - 2\cos x) = 0$$

$$\sin x = -1 \quad \text{OR} \quad \cos x = \frac{1}{2}$$
$$x = \frac{3\pi}{2} \qquad\qquad x = \frac{\pi}{3}, \frac{5\pi}{3}$$

Extraneous

The only solutions are $x = \dfrac{\pi}{3},\ \dfrac{5\pi}{3}$.

43. Use a calculator to find all solutions of $12\sin^2 x - 13\sin x + 3 = 0$ in the interval $[0,\ 2\pi)$.

Solution:

$$12\sin^2 x - 13\sin x + 3 = 0$$
$$(4\sin x - 3)(3\sin x - 1) = 0$$

$$\sin x = \frac{3}{4} \qquad\qquad \text{OR} \qquad \sin x = \frac{1}{3}$$
$$x = 0.8481,\ 2.2935 \qquad\qquad x = 0.3398,\ 2.8018$$

47. Use a calculator to find all solutions of $\tan^2 x - 8\tan x + 13 = 0$ in the interval $[0,\ 2\pi)$.

Solution:

$$\tan^2 x - 8\tan x + 13 = 0$$

$$\tan x = \frac{8 \pm \sqrt{64 - 52}}{2}$$

$$= \frac{8 \pm 2\sqrt{3}}{2}$$

$$\tan x = 4 + \sqrt{3} \qquad \text{OR} \qquad \tan x = 4 - \sqrt{3}$$

$$x = 1.3981,\ 4.5397 \qquad\qquad x = 1.1555,\ 4.2971$$

53. A 5-pound weight is oscillating on the end of a spring, and the position of the weight relative to the point of equilibrium is given by

$$h(t) = \frac{1}{4}(\cos 8t - 3\sin 8t)$$

where t is the time in seconds. Find the times when the weight is at the point of equilibrium $[h(t) = 0]$ for $0 \le t \le 1$.

Solution:

$$\frac{1}{4}(\cos 8t - 3\sin 8t) = 0$$

$$\cos 8t = 3\sin 8t$$

$$\frac{1}{3} = \tan 8t$$

$$8t = 0.32175 + n\pi$$

$$t = 0.04 + \frac{n\pi}{8}$$

In the interval $0 \le t \le 1$, we have $t = 0.04,\ 0.43,$ and 0.83.

SECTION 3.4

Sum and Difference Formulas

- You should memorize the sum and difference formulas given at the beginning of the section.

- You should be able to use these formulas to find the values of the trigonometric functions of angles whose sums or differences are special angles.

- You should be able to use these formulas to solve trigonometric equations.

Solutions to Selected Exercises

5. Determine the exact value of the sine, cosine, and tangent of the angle $195° = 225° - 30°$.

Solution:

$$\sin 195° = \sin(225° - 30°)$$
$$= \sin 225° \cos 30° - \cos 225° \sin 30°$$
$$= (-\sin 45°) \cos 30° - (-\cos 45°) \sin 30°$$
$$= -\frac{\sqrt{2}}{2} \cdot \frac{\sqrt{3}}{2} + \frac{\sqrt{2}}{2} \cdot \frac{1}{2} = \frac{\sqrt{2}}{4}(1 - \sqrt{3})$$

$$\cos 195° = \cos(225° - 30°)$$
$$= \cos 225° \cos 30° + \sin 225° \sin 30°$$
$$= (-\cos 45°) \cos 30° + (-\sin 45°) \sin 30°$$
$$= -\frac{\sqrt{2}}{2} \cdot \frac{\sqrt{3}}{2} - \frac{\sqrt{2}}{2} \cdot \frac{1}{2} = -\frac{\sqrt{2}}{4}(1 + \sqrt{3})$$

$$\tan 195° = \tan(225° - 30°) = \frac{\tan 225° - \tan 30°}{1 + \tan 225° \tan 30°} = \frac{\tan 45° - \tan 30°}{1 + \tan 45° \tan 30°}$$

$$= \frac{1 - \sqrt{3}/3}{1 + (1)(\sqrt{3}/3)} \cdot \frac{3}{3} = \frac{3 - \sqrt{3}}{3 + \sqrt{3}} \cdot \frac{3 - \sqrt{3}}{3 - \sqrt{3}}$$

$$= \frac{(3 - \sqrt{3})^2}{9 - 3} = \frac{[\sqrt{3}(\sqrt{3} - 1)]^2}{6}$$

$$= \frac{3(\sqrt{3} - 1)^2}{6} = \frac{1}{2}(\sqrt{3} - 1)^2$$

11. Simplify $\cos 25° \cos 15° - \sin 25° \sin 15°$.

Solution:

$$\cos 25° \cos 15° - \sin 25° \sin 15° = \cos(25° + 15°) = \cos 40°$$

15. Simplify

$$\frac{\tan 325° - \tan 86°}{1 + \tan 325° \tan 86°}.$$

Solution:

$$\frac{\tan 325° - \tan 86°}{1 + \tan 325° \tan 86°} = \tan(325° - 86°) = \tan 239°$$

19. Simplify

$$\frac{\tan 2x + \tan x}{1 - \tan 2x \tan x}.$$

Solution:

$$\frac{\tan 2x + \tan x}{1 - \tan 2x \tan x} = \tan(2x + x) = \tan 3x$$

23. Find the exact value of $\cos(v + u)$ given that $\sin u = \frac{5}{13}$, $0 < u < \pi/2$, and $\cos v = -\frac{3}{5}$, $\pi/2 < v < \pi$.

Solution:

$$\sin u = \frac{5}{13}, \ 0 < u < \frac{\pi}{2} \Rightarrow \cos u = \frac{12}{13}$$

$$\cos v = -\frac{3}{5}, \ \frac{\pi}{2} < v < \pi \Rightarrow \sin v = \frac{4}{5}$$

$$\cos(v + u) = \cos v \cos u - \sin v \sin u = \left(-\frac{3}{5}\right)\left(\frac{12}{13}\right) - \left(\frac{4}{5}\right)\left(\frac{5}{13}\right) = -\frac{56}{65}$$

27. Find the exact value of $\sin(v - u)$ given that $\sin u = \frac{7}{25}$, $\pi/2 < u < \pi$ and $\cos v = \frac{4}{5}$, $3\pi/2 < v < 2\pi$.

Solution:

$$\sin u = \frac{7}{25}, \ \frac{\pi}{2} < u < \pi \Rightarrow \cos u = -\frac{24}{25}$$

$$\cos v = \frac{4}{5}, \ \frac{3\pi}{2} < v < 2\pi \Rightarrow \sin v = -\frac{3}{5}$$

$$\sin(v - u) = \sin v \cos u - \cos v \sin u = \left(-\frac{3}{5}\right)\left(-\frac{24}{25}\right) - \left(\frac{4}{5}\right)\left(\frac{7}{25}\right) = \frac{44}{125}$$

31. Verify $\cos\left(\dfrac{3\pi}{2} - x\right) = -\sin x$.

Solution:

$$\cos\left(\frac{3\pi}{2} - x\right) = \cos\frac{3\pi}{2}\cos x + \sin\frac{3\pi}{2}\sin x = (0)(\cos x) + (-1)(\sin x) = -\sin x$$

35. Verify $\cos(\pi - \theta) + \sin\left(\dfrac{\pi}{2} + \theta\right) = 0$.

Solution:

$$\cos(\pi - \theta) + \sin\left(\frac{\pi}{2} + \theta\right) = (\cos\pi\cos\theta + \sin\pi\sin\theta) + \left(\sin\frac{\pi}{2}\cos\theta + \cos\frac{\pi}{2}\sin\theta\right)$$
$$= (-1)\cos\theta + (0)\sin\theta + (1)\cos\theta + (0)\sin\theta$$
$$= -\cos\theta + 0 + \cos\theta + 0$$
$$= 0$$

39. Verify $\cos(x + y)\cos(x - y) = \cos^2 x - \sin^2 y$.

Solution:

$$\cos(x + y)\cos(x - y) = (\cos x\cos y - \sin x\sin y)(\cos x\cos y + \sin x\sin y)$$
$$= \cos^2 x\cos^2 y - \sin^2 x\sin^2 y$$
$$= \cos^2 x(1 - \sin^2 y) - (1 - \cos^2 x)\sin^2 y$$
$$= \cos^2 x - \cos^2 x\sin^2 y - \sin^2 y + \cos^2 x\sin^2 y$$
$$= \cos^2 x - \sin^2 y$$

43. Verify

$$\sin(x + y + z) = \sin x\cos y\cos z + \sin y\cos x\cos z + \sin z\cos x\cos y - \sin x\sin y\sin z.$$

Solution:

$$\sin(x + y + z) = \sin[x + (y + z)]$$
$$= \sin x\cos(y + z) + \cos x\sin(y + z)$$
$$= \sin x(\cos y\cos z - \sin y\sin z) + \cos x(\sin y\cos z + \cos y\sin z)$$
$$= \sin x\cos y\cos z - \sin x\sin y\sin z + \sin y\cos x\cos z + \sin z\cos x\cos y$$
$$= \sin x\cos y\cos z + \sin y\cos x\cos z + \sin z\cos x\cos y - \sin x\sin y\sin z$$

47. Verify $a\sin B\theta + b\cos B\theta = \sqrt{a^2 + b^2}\sin(B\theta + C)$, where $C = \arctan(b/a)$.

Solution:

$$\sqrt{a^2 + b^2}\,\sin(B\theta + C) = \sqrt{a^2 + b^2}\,(\sin B\theta \cos C + \cos B\theta \sin C)$$

$$= \sqrt{a^2 + b^2}\left[\sin B\theta\left(\frac{a}{\sqrt{a^2 + b^2}}\right) + \cos B\theta\left(\frac{b}{\sqrt{a^2 + b^2}}\right)\right]$$

$$= \frac{\sqrt{a^2 + b^2}}{\sqrt{a^2 + b^2}}(a \sin B\theta + b \cos B\theta)$$

$$= a \sin B\theta + b \cos B\theta$$

51. Use the formulas given in Exercises 47 and 48 to write $12 \sin 3\theta + 5 \cos 3\theta$ in the following forms.

(a) $\sqrt{a^2 + b^2}\,\sin(B\theta + C)$ (b) $\sqrt{a^2 + b^2}\,\cos(B\theta - C)$

Solution:

$a = 12$, $b = 5$, $B = 3$, $C = \arctan \frac{5}{12} \approx 0.3948$

$\sqrt{12^2 + 5^2} = \sqrt{169} = 13$

(a) Thus, $\sqrt{a^2 + b^2}\,\sin(B\theta + C) = 13 \sin(3\theta + 0.3948)$, and

(b) $\sqrt{a^2 + b^2}\,\cos(B\theta - C) = 13 \cos(3\theta - 0.3948)$.

55. Write $\sin(\arcsin x + \arccos x)$ as an algebraic expression in x. [*Hint:* See Examples 6 and 7 in Section 7.7.]

Solution:

$$\sin(\arcsin x + \arccos x) = \sin(\arcsin x)\cos(\arccos x) + \cos(\arcsin x)\sin(\arccos x)$$

$$= (x)(x) + \frac{\sqrt{1 - x^2}}{1} \cdot \frac{\sqrt{1 - x^2}}{1} = x^2 + 1 - x^2 = 1$$

59. Find all solutions in the interval $[0,\ 2\pi)$ of $\cos\left(x + \dfrac{\pi}{4}\right) + \cos\left(x - \dfrac{\pi}{4}\right) = 1$.

Solution:

$$\cos\left(x + \frac{\pi}{4}\right) + \cos\left(x - \frac{\pi}{4}\right) = 1$$

$$\cos x \cos \frac{\pi}{4} - \sin x \sin \frac{\pi}{4} + \cos x \cos \frac{\pi}{4} + \sin x \sin \frac{\pi}{4} = 1$$

$$\frac{\sqrt{2}}{2}\cos x + \frac{\sqrt{2}}{2}\cos x = 1$$

$$\sqrt{2}\cos x = 1$$

$$\cos x = \frac{1}{\sqrt{2}}$$

$$x = \frac{\pi}{4},\ \frac{7\pi}{4}$$

63. Show that

$$\frac{\cos(x+h) - \cos x}{h} = \cos x \left(\frac{\cos h - 1}{h}\right) - \sin x \left(\frac{\sin h}{h}\right).$$

Solution:

$$\frac{\cos(x+h) - \cos x}{h} = \frac{\cos x \cos h - \sin x \sin h - \cos x}{h}$$

$$= \frac{\cos x(\cos h - 1) - \sin x \sin h}{h}$$

$$= \frac{\cos x(\cos h - 1)}{h} - \frac{\sin x \sin h}{h}$$

$$= \cos x \left(\frac{\cos h - 1}{h}\right) - \sin x \left(\frac{\sin h}{h}\right)$$

SECTION 3.5

Multiple-Angle Formulas and Product-Sum Formulas

■ You should know the following double-angle formulas.

(a) $\sin 2u = 2 \sin u \cos u$

(b) $\cos 2u = \cos^2 u - \sin^2 u$

$$= 2 \cos^2 u - 1$$

$$= 1 - 2 \sin^2 u$$

(c) $\tan 2u = \dfrac{2 \tan u}{1 - \tan^2 u}$

■ You should be able to reduce the power of a trigonometric function.

(a) $\sin^2 u = \dfrac{1 - \cos 2u}{2}$

(b) $\cos^2 u = \dfrac{1 + \cos 2u}{2}$

(c) $\tan^2 u = \dfrac{1 - \cos 2u}{1 + \cos 2u}$

■ You should be able to use the half-angle formulas.

(a) $\sin^2 \dfrac{u}{2} = \dfrac{1 - \cos u}{2}$

(b) $\cos^2 \dfrac{u}{2} = \dfrac{1 + \cos u}{2}$

(c) $\tan^2 \dfrac{u}{2} = \dfrac{1 - \cos u}{1 + \cos u}$

■ You should be able to use the following.

Product-Sum Formulas

(a) $\sin u \sin v = \dfrac{1}{2}[\cos(u - v) - \cos(u + v)]$

(b) $\cos u \cos v = \dfrac{1}{2}[\cos(u - v) + \cos(u + v)]$

(c) $\sin u \cos v = \dfrac{1}{2}[\sin(u + v) + \sin(u - v)]$

(d) $\cos u \sin v = \dfrac{1}{2}[\sin(u + v) - \sin(u - v)]$

■ You should be able to use the following.

Sum-Product Formulas

(a) $\sin x + \sin y = 2 \sin\left(\dfrac{x+y}{2}\right) \cos\left(\dfrac{x-y}{2}\right)$

(b) $\sin x - \sin y = 2 \cos\left(\dfrac{x+y}{2}\right) \sin\left(\dfrac{x-y}{2}\right)$

(c) $\cos x + \cos y = 2 \cos\left(\dfrac{x+y}{2}\right) \cos\left(\dfrac{x-y}{2}\right)$

(d) $\cos x - \cos y = -2 \sin\left(\dfrac{x+y}{2}\right) \sin\left(\dfrac{x-y}{2}\right)$

Solutions to Selected Exercises

3. Use a double-angle formula to determine the exact values of the sine, cosine, and tangent of the angle $60° = 2(30°)$.

Solution:

$$\sin 60° = 2 \sin 30° \cos 30° = 2\left(\frac{1}{2}\right)\left(\frac{\sqrt{3}}{2}\right) = \frac{\sqrt{3}}{2}$$

$$\cos 60° = \cos^2 30° - \sin^2 30° = \left(\frac{\sqrt{3}}{2}\right)^2 - \left(\frac{1}{2}\right)^2 = \frac{1}{2}$$

$$\tan 60° = \frac{2 \tan 30°}{1 - \tan^2 30°} = \frac{2(1/\sqrt{3})}{1 - (1/\sqrt{3})^2} = \frac{2/\sqrt{3}}{2/3} = \frac{3}{\sqrt{3}} = \sqrt{3}$$

9. Find the exact values of $\sin 2u$, $\cos 2u$, and $\tan 2u$, given $\tan u = \frac{1}{2}$, $\pi < u < 3\pi/2$.

Solution:

$$\tan u = \frac{1}{2}, \quad u \text{ lies in Quadrant III}$$

$$\sin u = -\frac{1}{5} \text{ and } \cos u = -\frac{2}{\sqrt{5}}$$

$$\sin 2u = 2\left(-\frac{1}{\sqrt{5}}\right)\left(-\frac{2}{\sqrt{5}}\right) = \frac{4}{5}$$

$$\cos 2u = \left(-\frac{2}{\sqrt{5}}\right)^2 - \left(-\frac{1}{\sqrt{5}}\right)^2 = \frac{3}{5}$$

$$\tan 2u = \frac{\sin 2u}{\cos 2u} = \frac{4}{3}$$

15. Use half-angle formulas to determine the exact values of the sine, cosine, and tangent of the angle $112°30' = \frac{1}{2}(225°)$.

Solution:

$$\sin 112°30' = +\sqrt{\frac{1-\cos 225°}{2}} = \sqrt{\frac{1-(-\sqrt{2}/2)}{2}} = \sqrt{\frac{2+\sqrt{2}}{4}} = \frac{\sqrt{2+\sqrt{2}}}{2}$$

$$\cos 112°30' = -\sqrt{\frac{1+\cos 225°}{2}} = -\sqrt{\frac{1+(-\sqrt{2}/2)}{2}} = -\sqrt{\frac{2-\sqrt{2}}{4}} = -\frac{\sqrt{2-\sqrt{2}}}{2}$$

$$\tan 112°30' = -\sqrt{\frac{1-\cos 225°}{1+\cos 225°}} = -\sqrt{\frac{1+\sqrt{2}/2}{1-\sqrt{2}/2}} = -\sqrt{\frac{2+\sqrt{2}}{2-\sqrt{2}}} \cdot \sqrt{\frac{2+\sqrt{2}}{2+\sqrt{2}}}$$

$$= -\sqrt{\frac{(2+\sqrt{2})^2}{4-2}} = -\frac{2+\sqrt{2}}{\sqrt{2}} = -(\sqrt{2}+1) = -1-\sqrt{2}$$

21. Find the exact values of $\sin(u/2)$, $\cos(u/2)$, and $\tan(u/2)$ by using the half-angle formulas, given $\tan u = -\frac{5}{8}$, $3\pi/2 < u < 2\pi$.

Solution:

$$\tan u = -\frac{5}{8}, \ u \text{ lies in Quadrant IV}$$

$$\sin u = -\frac{5}{\sqrt{89}} \text{ and } \cos u = \frac{8}{\sqrt{89}}$$

$$\sin \frac{u}{2} = \sqrt{\frac{1-8/\sqrt{89}}{2}} = \sqrt{\frac{\sqrt{89}-8}{2\sqrt{89}}}$$

$$\cos \frac{u}{2} = -\sqrt{\frac{1+8/\sqrt{89}}{2}} = -\sqrt{\frac{\sqrt{89}+8}{2\sqrt{89}}}$$

$$\tan \frac{u}{2} = -\sqrt{\frac{1-8/\sqrt{89}}{1+8/\sqrt{89}}} = -\sqrt{\frac{\sqrt{89}-8}{\sqrt{89}+8}} \cdot \sqrt{\frac{\sqrt{89}-8}{\sqrt{89}-8}}$$

$$= \sqrt{\frac{(\sqrt{89}-8)^2}{89-64}} = -\left(\frac{\sqrt{89}-8}{5}\right) = \frac{1}{5}(8-\sqrt{89})$$

25. Use the half-angle formulas to simplify

$$\sqrt{\frac{1 - \cos 6x}{2}}.$$

Solution:

$$\sqrt{\frac{1 - \cos 6x}{2}} = \sqrt{\frac{1 - \cos 2(3x)}{2}} = \sin 3x$$

29. Use the power-reducing formulas to write each expression in terms of the first power of the cosine.

(a) $\cos^4 x$ (b) $\sin^2 x \cos^4 x$

Solution:

(a) $\cos^4 x = (\cos^2 x)^2$

$$= \left(\frac{1 + \cos 2x}{2}\right)^2$$

$$= \frac{1}{4}(1 + 2\cos 2x + \cos^2 2x)$$

$$= \frac{1}{4}\left(1 + 2\cos 2x + \frac{1 + \cos 4x}{2}\right)$$

$$= \frac{1}{4}\left(\frac{3}{2} + 2\cos 2x + \frac{1}{2}\cos 4x\right)$$

$$= \frac{1}{8}(3 + 4\cos 2x + \cos 4x)$$

(b) $\sin^2 x \cos^4 x = \left(\frac{1 - \cos 2x}{2}\right)\cos^4 x$

$$= \frac{1}{2}(1 - \cos 2x)\frac{1}{8}(3 + 4\cos 2x + \cos 4x) \text{ from part (a)}$$

$$= \frac{1}{16}(1 - \cos 2x)(3 + 4\cos 2x + \cos 4x)$$

33. Rewrite $\sin 5\theta \cos 3\theta$ as a sum.

Solution:

$$\sin 5\theta \cos 3\theta = \frac{1}{2}[\sin(5\theta + 3\theta) + \sin(5\theta - 3\theta)] = \frac{1}{2}(\sin 8\theta + \sin 2\theta)$$

37. Rewrite $\sin(x + y)\sin(x - y)$ as a sum.

Solution:

$$\sin(x + y)\sin(x - y) = \frac{1}{2}\{\cos[(x + y) - (x - y)] - \cos[(x + y) + (x - y)]\}$$
$$= \frac{1}{2}\{\cos 2y - \cos 2x\}$$

41. Express $\sin 60° + \sin 30°$ as a product.

Solution:

$$\sin 60° + \sin 30° = 2\sin\left(\frac{60° + 30°}{2}\right)\cos\left(\frac{60° - 30°}{2}\right) = 2\sin 45° \cos 15°$$

45. Express $\cos 6x + \cos 2x$ as a product.

Solution:

$$\cos 6x + \cos 2x = 2\cos\left(\frac{6x + 2x}{2}\right)\cos\left(\frac{6x - 2x}{2}\right) = 2\cos 4x \cos 2x$$

49. Express $\cos(\phi + 2\pi) + \cos\phi$ as a product.

Solution:

$$\cos(\phi + 2\pi) + \cos\phi = 2\cos\left(\frac{\phi + 2\pi + \phi}{2}\right)\cos\left(\frac{\phi + 2\pi - \phi}{2}\right) = 2\cos(\phi + \pi)\cos\pi$$

51. Verify

$$\csc 2\theta = \frac{\csc\theta}{2\cos\theta}.$$

Solution:

$$\csc 2\theta = \frac{1}{\sin 2\theta} = \frac{1}{2\sin\theta(\cos\theta)} = \frac{1}{\sin\theta} \cdot \frac{1}{2\cos\theta} = \frac{\csc\theta}{2\cos\theta}$$

55. Verify $(\sin x + \cos x)^2 = 1 + \sin 2x$.

Solution:

$$
\begin{aligned}
(\sin x + \cos x)^2 &= \sin^2 x + 2\sin x \cos x + \cos^2 x \\
&= (\sin^2 x + \cos^2 x) + 2\sin x \cos x \\
&= 1 + \sin 2x
\end{aligned}
$$

59. Verify $1 + \cos 10y = 2\cos^2 5y$.

Solution:

$$2\cos^2 5y = 2\left[\frac{1 + \cos 10y}{2}\right] = 1 + \cos 10y$$

65. Verify

$$\frac{\cos 4x - \cos 2x}{2 \sin 3x} = -\sin x.$$

Solution:

$$\frac{\cos 4x - \cos 2x}{2 \sin 3x} = \frac{-2 \sin\left(\dfrac{4x + 2x}{2}\right) \sin\left(\dfrac{4x - 2x}{2}\right)}{2 \sin 3x} = \frac{-2 \sin 3x \sin x}{2 \sin 3x} = -\sin x$$

69. Verify

$$\frac{\cos t + \cos 3t}{\sin 3t - \sin t} = \cot t.$$

Solution:

$$\frac{\cos t + \cos 3t}{\sin 3t - \sin t} = \frac{2 \cos\left(\dfrac{t + 3t}{2}\right) \cos\left(\dfrac{t - 3t}{2}\right)}{2 \cos\left(\dfrac{3t + t}{2}\right) \sin\left(\dfrac{3t - t}{2}\right)}$$

$$= \frac{2 \cos 2t \cos(-t)}{2 \cos 2t \sin t} = \frac{\cos t}{\sin t} = \cot t$$

73. Find all the solutions of $4 \sin x \cos x = 1$ in the interval $[0, \ 2\pi)$.

Solution:

$$4 \sin x (\cos x) = 1$$
$$2[2(\sin x) \cos x] = 1$$
$$2 \sin 2x = 1$$
$$\sin 2x = \frac{1}{2}$$

$$2x = \frac{\pi}{6}, \quad 2x = \frac{5\pi}{6}, \quad 2x = \frac{13\pi}{6}, \quad 2x = \frac{17\pi}{6}$$

$$x = \frac{\pi}{12}, \quad x = \frac{5\pi}{12}, \quad x = \frac{13\pi}{12}, \quad x = \frac{17\pi}{12}$$

77. Find all solutions of $\sin 4x + 2\sin 2x = 0$ in the interval $[0,\ 2\pi)$.

Solution:

$$\sin 4x + 2\sin 2x = 0$$
$$2\sin 2x \cos 2x + 2\sin 2x = 0$$
$$2\sin 2x(\cos 2x + 1) = 0$$

$\sin 2x = 0$	or $\quad \cos 2x = -1$
$2x = 0,\ \pi,\ 2\pi,\ 3\pi$	$2x = \pi,\ 3\pi$
$x = 0,\ \dfrac{\pi}{2},\ \pi,\ \dfrac{3\pi}{2}$	$x = \dfrac{\pi}{2},\ \dfrac{3\pi}{2}$

81. Find all solutions of $\sin 6x + \sin 2x = 0$ in the interval $[0,\ 2\pi)$.

Solution:

$$\sin 6x + \sin 2x = 0$$
$$2\sin\left(\frac{6x + 2x}{2}\right)\cos\left(\frac{6x - 2x}{2}\right) = 0$$
$$2(\sin 4x)\cos 2x = 0$$

$\sin 4x = 0$	or $\quad \cos 2x = 0$
$4x = n\pi$	$2x = \dfrac{\pi}{2} + n\pi$
$x = \dfrac{n\pi}{4}$	$x = \dfrac{\pi}{4} + \dfrac{n\pi}{2}$

In the interval $[0,\ 2\pi)$, we have

$$x = 0,\ \frac{\pi}{4},\ \frac{\pi}{2},\ \frac{3\pi}{4},\ \pi,\ \frac{5\pi}{4},\ \frac{3\pi}{2},\ \frac{7\pi}{4}.$$

85. Sketch the graph of $f(x) = \sin^2 x$ by using the power-reducing formulas.

Solution:

$$f(x) = \sin^2 x$$
$$= \frac{1 - \cos 2x}{2}$$
$$= \frac{1}{2} - \frac{1}{2}\cos 2x$$

Period: π

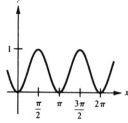

x	$-\pi$	$-\dfrac{3\pi}{4}$	$-\dfrac{\pi}{2}$	$-\dfrac{\pi}{4}$	0	$\dfrac{\pi}{4}$	$\dfrac{\pi}{2}$	$\dfrac{3\pi}{4}$	π
$f(x)$	0	$\dfrac{1}{2}$	1	$\dfrac{1}{2}$	0	$\dfrac{1}{2}$	1	$\dfrac{1}{2}$	0

89. Verify $\sin(\phi - \theta) = \cos 2\theta$ for complementary angles ϕ and θ.

Solution:

Since ϕ and θ are complementary, $\phi = 90° - \theta$.

$$\begin{aligned} \sin(\phi - \theta) &= \sin[(90° - \theta) - \theta] \\ &= \sin[90° - 2\theta] \\ &= \sin 90° \cos 2\theta - \cos 90° \sin 2\theta \\ &= (1) \cos 2\theta - (0) \sin 2\theta \\ &= \cos 2\theta \end{aligned}$$

93. Prove $\cos u \sin v = \frac{1}{2}[\sin(u + v) - \sin(u - v)]$.

Solution:

$$\begin{aligned} \frac{1}{2}[\sin(u + v) - \sin(u - v)] &= \frac{1}{2}\{(\sin u \cos v + \cos u \sin v) - (\sin u \cos v - \cos u \sin v)\} \\ &= \frac{1}{2}\{2 \cos u \sin v\} \\ &= \cos u \sin v \end{aligned}$$

REVIEW EXERCISES FOR CHAPTER 3

Solutions to Selected Exercises

3. Simplify

$$\frac{\sin^2 \alpha - \cos^2 \alpha}{\sin^2 \alpha - \sin \alpha \cos \alpha}.$$

Solution:

$$\frac{\sin^2 \alpha - \cos^2 \alpha}{\sin^2 \alpha - \sin \alpha \cos \alpha} = \frac{(\sin \alpha + \cos \alpha)(\sin \alpha - \cos \alpha)}{\sin \alpha (\sin \alpha - \cos \alpha)}$$

$$= \frac{\sin \alpha + \cos \alpha}{\sin \alpha} = \frac{\sin \alpha}{\sin \alpha} + \frac{\cos \alpha}{\sin \alpha} = 1 + \cot \alpha$$

7. Simplify $\tan^2 \theta (\csc^2 \theta - 1)$.

Solution:

$$\tan^2 \theta (\csc^2 \theta - 1) = \tan^2 \theta (\cot^2 \theta) = \tan^2 \theta \left(\frac{1}{\tan^2 \theta}\right) = 1$$

11. Verify $\tan x (1 - \sin^2 x) = \frac{1}{2} \sin 2x$.

Solution:

$$\tan x (1 - \sin^2 x) = \frac{\sin x}{\cos x}(\cos^2 x) = (\sin x) \cos x = \frac{1}{2}(2(\sin x) \cos x) = \frac{1}{2} \sin 2x$$

15. Verify $\sin^2 x \cos^4 x = \frac{1}{16}(1 - \cos 4x + 2\sin^2 2x \cos 2x)$.

Solution:

$$\frac{1}{16}(1 - \cos 4x + 2\sin^2 2x \cos 2x) = \frac{1}{16}[1 - (\cos^2 2x - \sin^2 2x) + 2(\sin^2 2x)\cos 2x]$$

$$= \frac{1}{16}[2\sin^2 2x + 2(\sin^2 2x)\cos 2x]$$

$$= \frac{1}{8}[\sin^2 2x(1 + \cos 2x)]$$

$$= \frac{1}{8}[(2(\sin x)\cos x)^2(1 + \cos^2 x - \sin^2 x)]$$

$$= \frac{1}{8}[4(\sin^2 x)\cos^2 x(2\cos^2 x)]$$

$$= \sin^2 x(\cos^4 x)$$

19. Verify $\sqrt{\dfrac{1 - \sin \theta}{1 + \sin \theta}} = \dfrac{1 - \sin \theta}{|\cos \theta|}$.

Solution:

$$\sqrt{\frac{1 - \sin\theta}{1 + \sin\theta}} = \sqrt{\frac{1 - \sin\theta}{1 + \sin\theta} \cdot \frac{1 - \sin\theta}{1 - \sin\theta}} = \sqrt{\frac{(1 - \sin\theta)^2}{\cos^2\theta}} = \frac{1 - \sin\theta}{|\cos\theta|}$$

23. Verify $\sin\left(x - \dfrac{3\pi}{2}\right) = \cos x$.

Solution:

$$\sin\left(x - \frac{3\pi}{2}\right) = (\sin x)\cos\frac{3\pi}{2} - (\cos x)\sin\frac{3\pi}{2} = (\sin x)(0) - \cos x(-1) = \cos x$$

27. Verify $\dfrac{\cos 3x - \cos x}{\sin 3x - \sin x} = -\tan 2x$.

Solution:

$$\frac{\cos 3x - \cos x}{\sin 3x - \sin x} = \frac{-2\sin\left(\dfrac{3x + x}{2}\right)\sin\left(\dfrac{3x - x}{2}\right)}{2\cos\left(\dfrac{3x + x}{2}\right)\sin\left(\dfrac{3x - x}{2}\right)} = \frac{-2(\sin 2x)\sin x}{2(\cos 2x)\sin x} = -\frac{\sin 2x}{\cos 2x} = -\tan 2x$$

31. Verify $2\sin y\cos y\sec 2y = \tan 2y$.

Solution:

$$2\sin y\cos y\sec 2y = \sin 2y\sec 2y = \sin 2y\left(\frac{1}{\cos 2y}\right) = \frac{\sin 2y}{\cos 2y} = \tan 2y$$

35. Verify $\tan^2 x = \dfrac{1 - \cos 2x}{1 + \cos 2x}$.

Solution:

$$\frac{1 - \cos 2x}{1 + \cos 2x} = \frac{1 - \left(1 - 2\sin^2 x\right)}{1 + \left(2\cos^2 x - 1\right)} = \frac{2\sin^2 x}{2\cos^2 x} = \tan^2 x$$

39. Verify $1 + \cos 2x + \cos 4x + \cos 6x = 4\cos x\cos 2x\cos 3x$.

Solution:

$$4\cos x\cos 2x\cos 3x = 4\cos x\left[\tfrac{1}{2}(\cos(-x) + \cos(5x))\right] = 2\cos x(\cos x + \cos 5x)$$
$$= 2\cos^2 x + 2\cos x\cos 5x = 2\cos^2 x + 2\left[\tfrac{1}{2}(\cos(-4x) + \cos 6x)\right]$$
$$= 2\cos^2 x + \cos(-4x) + \cos 6x = 1 + \cos 2x + \cos 4x + \cos 6x$$

43. Using the sum, difference, or half-angle formulas, find the exact value of

$$\cos(157°\,30') = \cos\frac{315°}{2}.$$

Solution:

$$\cos(157°\,30') = \cos\frac{315°}{2} = -\sqrt{\frac{1+\cos 315°}{2}} = -\sqrt{\frac{1+\cos 45°}{2}}$$

$$= -\sqrt{\frac{1+\frac{\sqrt{2}}{2}}{2}} = -\sqrt{\frac{2+\sqrt{2}}{4}} = -\frac{\sqrt{2+\sqrt{2}}}{2}$$

47. Find the exact value of $\cos(u-v)$, given that $\sin u = \frac{3}{4}$, $\cos v = -\frac{5}{13}$, (u and v are in Quadrant II).

Solution:

Since u and v are in Quadrant II,

$$\sin u = \frac{3}{4} \Rightarrow \cos u = -\frac{\sqrt{7}}{4}$$

$$\cos v = -\frac{5}{13} \Rightarrow \sin v = \frac{12}{13}.$$

$$\cos(u-v) = \cos u \cos v + \sin u \sin v = \left(-\frac{\sqrt{7}}{4}\right)\left(-\frac{5}{13}\right) + \left(\frac{3}{4}\right)\left(\frac{12}{13}\right) = \frac{5\sqrt{7}+36}{52}$$

51. Find all solutions of $\sin x - \tan x = 0$ in the interval $[0,\ 2\pi)$.

Solution:

$$\sin x - \tan x = 0$$

$$\sin x - \frac{\sin x}{\cos x} = 0$$

$$(\sin x)\cos x - \sin x = 0$$

$$\sin x(\cos x - 1) = 0$$

$$\sin x = 0 \qquad \text{OR} \qquad \cos x = 1$$

$$x = 0,\ \pi \qquad\qquad\qquad x = 0$$

55. Find all solutions of $\sin 2x + \sqrt{2}\sin x = 0$ in the interval $[0,\ 2\pi)$.

Solution:

$$\sin 2x + \sqrt{2}\sin x = 0$$

$$2(\sin x)\cos x + \sqrt{2}\sin x = 0$$

$$\sin x(2\cos x + \sqrt{2}) = 0$$

$$\sin x = 0 \qquad \text{OR} \quad \cos x = -\frac{\sqrt{2}}{2}$$

$$x = 0,\ \pi \qquad\qquad\qquad x = \frac{3\pi}{4},\ \frac{5\pi}{4}$$

59. Find all solutions of $\tan^3 x - \tan^2 x + 3\tan x - 3 = 0$ in the interval $[0,\ 2\pi)$.

Solution:

$$\tan^3 x - \tan^2 x + 3\tan x - 3 = 0$$
$$(\tan x - 1)(\tan^2 x + 3) = 0$$

$\tan x = 1$	OR	$\tan^2 x = -3$
$x = \dfrac{\pi}{4},\ \dfrac{5\pi}{4}$		No real solutions

63. Write $\sin 3\alpha \sin 2\alpha$ as a sum or difference.

Solution:

$$\sin 3\alpha \sin 2\alpha = \tfrac{1}{2}[\cos(3\alpha - 2\alpha) - \cos(3\alpha + 2\alpha)] = \tfrac{1}{2}[\cos \alpha - \cos 5\alpha]$$

65. A standing wave on a string of given length is modeled by the equation

$$y = A\left(\cos\left[2\pi\left(\frac{t}{T} - \frac{x}{\lambda}\right)\right] + \cos\left[2\pi\left(\frac{t}{T} + \frac{x}{\lambda}\right)\right]\right).$$

Use the trigonometric identities for the cosine of the sum and difference of two angles to verify that the following equation is an equivalent model for the standing wave.

$$y = 2A\cos\frac{2\pi t}{T}\cos\frac{2\pi x}{\lambda}$$

Solution:

$$y = A\left(\cos\left[2\pi\left(\frac{t}{T} - \frac{x}{\lambda}\right)\right] + \cos\left[2\pi\left(\frac{t}{T} + \frac{x}{\lambda}\right)\right]\right)$$

$$= A\left(2\cos\left[\frac{2\pi\left(\frac{t}{T} - \frac{x}{\lambda}\right) + 2\pi\left(\frac{t}{T} + \frac{x}{\lambda}\right)}{2}\right]\cos\left[\frac{2\pi\left(\frac{t}{T} - \frac{x}{\lambda}\right) - 2\pi\left(\frac{t}{T} + \frac{x}{\lambda}\right)}{2}\right]\right)$$

$$= 2A\cos\left(\frac{4\pi\left(\frac{t}{T}\right)}{2}\right)\cos\left(\frac{-4\pi\left(\frac{x}{\lambda}\right)}{2}\right)$$

$$= 2A\cos\left(\frac{2\pi t}{T}\right)\cos\left(-\frac{2\pi x}{\lambda}\right)$$

$$= 2A\cos\left(\frac{2\pi t}{T}\right)\cos\left(\frac{2\pi x}{\lambda}\right) \quad \text{since} \quad \cos(-\theta) = \cos\theta.$$

Practice Test for Chapter 3

1. Find the value of the other five trigonometric functions, given $\tan x = \frac{4}{11}$, $\sec x < 0$.

2. Simplify $\dfrac{\sec^2 x + \csc^2 x}{\csc^2 x \left(1 + \tan^2 x\right)}$.

3. Rewrite as a single logarithm and simplify $\ln |\tan \theta| - \ln |\cot \theta|$.

4. True or False: $\cos\left(\dfrac{\pi}{2} - x\right) = \dfrac{1}{\csc x}$?

5. Factor and simplify $\sin^4 x + \left(\sin^2 x\right)\cos^2 x$.

6. Multiply and simplify $(\csc x + 1)(\csc x - 1)$.

7. Rationalize the denominator and simplify $\dfrac{\cos^2 x}{1 - \sin x}$.

8. Verify $\dfrac{1 + \cos \theta}{\sin \theta} + \dfrac{\sin \theta}{1 + \cos \theta} = 2 \csc \theta$.

9. Verify $\tan^4 x + 2\tan^2 x + 1 = \sec^4 x$.

10. Use the sum or difference formulas to determine:
    (a) $\sin 105°$
    (b) $\tan 15°$

11. Simplify $\left(\sin 42°\right)\cos 38° - \left(\cos 42°\right)\sin 38°$.

12. Verify $\tan\left(\theta + \dfrac{\pi}{4}\right) = \dfrac{1 + \tan \theta}{1 - \tan \theta}$.

13. Write $\sin(\arcsin x - \arccos x)$ as an algebraic expression in x.

14. Use the double-angle formulas to determine:
    (a) $\cos 120°$
    (b) $\tan 300°$

15. Use the half-angle formulas to determine:
    (a) $\sin 22.5°$
    (b) $\tan \dfrac{\pi}{12}$

16. Given $\sin = \dfrac{4}{5}$, θ lies in Quadrant II, find $\cos \dfrac{\theta}{2}$.

17. Use the power-reducing identities to write $(\sin^2 x) \cos^2 x$ in terms of the first power of cosine.

18. Rewrite as a sum $6(\sin 5\theta) \cos 2\theta$.

19. Rewrite as a product $\sin(x + \pi) + \sin(x - \pi)$.

20. Verify $\dfrac{\sin 9x + \sin 5x}{\cos 9x - \cos 5x} = -\cot 2x$.

21. Verify $(\cos u) \sin v = \frac{1}{2}[\sin(u + v) - \sin(u - v)]$.

22. Find all solutions in the interval $[0, \ 2\pi)$ $\ 4\sin^2 x = 1$.

23. Find all solutions in the interval $[0, \ 2\pi)$ $\ \tan^2 \theta + (\sqrt{3} - 1) \tan \theta - \sqrt{3} = 0$.

24. Find all solutions in the interval $[0, \ 2\pi)$ $\ \sin 2x = \cos x$.

25. Use the quadratic formula to find all solutions in the interval $[0, \ 2\pi)$ $\ \tan^2 x - 6\tan + 4 = 0$.

CHAPTER 4

Additional Applications of Trigonometry

SECTION 4.1

Law of Sines

■ If ABC is any oblique triangle with sides a, b, and c, then

$$\frac{a}{\sin A} = \frac{b}{\sin B} = \frac{c}{\sin C}.$$

■ You should be able to use the Law of Sines to solve an oblique triangle for the remaining three parts, given:

(a) Two angles and any side (AAS or ASA)
(b) Two sides and an angle opposite one of them (SSA)
    1. If A is acute and:
        (a) $a < h$, no triangle is possible.
        (b) $a = h$ or $a > b$, one triangle is possible.
        (c) $h < a < b$, two triangles are possible.
    2. If A is obtuse and:
        (a) $a \le b$, no triangle is possible.
        (b) $a > b$, one triangle is possible.

■ The area of any triangle equals one-half the product of the lengths of two sides times the sine of their included angle.

$$A = \frac{1}{2}ab\sin C = \frac{1}{2}ac\sin B = \frac{1}{2}bc\sin A$$

Solutions to Selected Exercises

3. Find the remaining sides and angles of the triangle.

Solution:
$A = 10°$, $B = 60°$, $a = 4.5$
$C = 180° - (10° + 60°) = 110°$

$$\frac{4.5}{\sin 10°} = \frac{b}{\sin 60°}$$
$$b = \sin 60° \left(\frac{4.5}{\sin 10°}\right) \approx 22.4$$
$$\frac{4.5}{\sin 10°} = \frac{c}{\sin 110°}$$
$$c = \sin 110° \left(\frac{4.5}{\sin 10°}\right) \approx 24.4$$

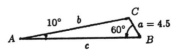

7. Find the remaining sides and angles of the triangle, given $A = 150°$, $C = 20°$, $a = 200$.

Solution:

$A = 150°$, $C = 20°$, $a = 200$

$B = 180° - (150° + 20°) = 10°$

$$\frac{200}{\sin 150°} = \frac{b}{\sin 10°}$$

$$b = \sin 10° \left(\frac{200}{\sin 150°}\right) \approx 69.5$$

$$\frac{200}{\sin 150°} = \frac{c}{\sin 20°}$$

$$c = \sin 20° \left(\frac{200}{\sin 150°}\right) \approx 136.8$$

11. Find the remaining sides and angles of the triangle, given $B = 15°30'$, $a = 4.5$, $b = 6.8$.

Solution:

$B = 15°30'$, $a = 4.5$, $b = 6.8$

$B = 15°30' = 15.5°$

$$\frac{6.8}{\sin 15.5°} = \frac{4.5}{\sin A}$$

$$\sin A = \frac{4.5 \sin 15.5}{6.8}$$

$$\sin A = 0.1768$$

$$A \approx 10.19° \approx 10°11'$$

$$C = 180° - (10°11' + 15°30') = 154°19'$$

$$\frac{6.8}{\sin 15.5°} = \frac{c}{\sin 154.32°}$$

$$c \approx 11$$

15. Find the remaining sides and angles of the triangle, given $A = 110°15'$, $a = 48$, $b = 16$.

Solution:

$A = 100°15'$, $a = 48$, $b = 16$

$$\frac{48}{\sin 100.25°} = \frac{16}{\sin B}$$

$$\sin B = 0.3127$$

$$B \approx 18.22° = 18°13'$$

$$C = 180° - (110°15' + 18°13') = 51°32'$$

$$\frac{48}{\sin 110.25°} = \frac{c}{\sin 51.53°}$$

$$c \approx 40.1$$

19. Solve the triangle: $A = 58°$, $a = 4.5$, $b = 5$, (if possible). If two solutions exist, find both.

Solution:

$A = 58°$, $a = 4.5$, $b = 5$

$h = b \sin A = 5 \sin 58° = 4.24$

A is acute and $h < a < b$, so there are two possible solutions.

$$\frac{4.5}{\sin 58°} = \frac{5}{\sin B}$$

$$\sin B = 0.9423$$

$$B \approx 70.4° \quad \text{or} \quad B \approx 109.6°$$

$$C = 51.6° \qquad\qquad C = 12.4°$$

$$\frac{4.5}{\sin 58°} = \frac{c}{\sin 51.6°} \qquad \frac{4.5}{\sin 58°} = \frac{c}{\sin 12.4°}$$

$$c \approx 4.16 \qquad\qquad c \approx 1.14$$

25. Find the area of the triangle with $C = 120°$, $a = 4$, and $b = 6$.

Solution:

$$\text{Area} = \tfrac{1}{2}ab \sin C = \tfrac{1}{2}(4)(6) \sin 120° = 10.4 \text{ square units}$$

31. Find the length d of the brace required to support the streetlight shown in the figure.

Solution:

$$\frac{3}{\sin 30°} = \frac{5}{\sin \alpha}$$

$$\sin \alpha = 0.8333$$

$$\alpha \approx 56.44°$$

$$\beta = 93.56°$$

$$\frac{3}{\sin 30°} = \frac{d}{\sin 93.56°}$$

$$d \approx 6 \text{ feet}$$

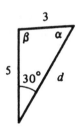

37. Two fire towers A and B are 18.5 miles apart. The bearing from A to B is N 65° E. A fire is spotted by the ranger in each tower, and its bearings from A and B are N 28° E and N 16.5° W, respectively (see figure). Find the distance of the fire from each tower.

Solution:

$A = 37°$, $B = 98.5°$, $C = 44.5°$, $c = 18.5$

$$\frac{b}{\sin 98.5°} = \frac{18.5}{\sin 44.5°}$$
$$b = 26.1 \text{ mi}$$
$$\frac{a}{\sin 37°} = \frac{18.5}{\sin 44.5°}$$
$$a = 15.9 \text{ mi}$$

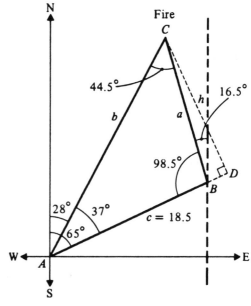

41. The following information about a triangular parcel of land is given at a zoning board meeting: "One side is 450 feet long and another is 120 feet long. The angle opposite the shorter side is 30°." Could this information be correct?

Solution:

$$h = b \sin A = 450 \sin 30° = 225$$
$$a = 120$$

Since $a < h$, no such triangle is possible.

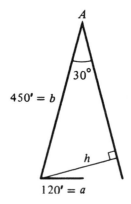

SECTION 4.2

Law of Cosines

■ If ABC is any oblique triangle with sides a, b, and c, then the following equations are valid.

(a) $a^2 = b^2 + c^2 - 2bc \cos A$ or $\cos A = \dfrac{b^2 + c^2 - a^2}{2bc}$

(b) $b^2 = a^2 + c^2 - 2ac \cos B$ or $\cos B = \dfrac{a^2 + c^2 - b^2}{2ac}$

(c) $c^2 = a^2 + b^2 - 2ab \cos C$ or $\cos C = \dfrac{a^2 + b^2 - c^2}{2ab}$

■ You should be able to use the Law of Cosines to solve an oblique triangle for the remaining three parts, given:

(a) Three sides (SSS)
(b) Two sides and their included angle (SAS)

■ Given any triangle with sides of length a, b, and c, then the area of the triangle is

$$\text{Area} = \sqrt{s(s-a)(s-b)(s-c)}, \text{ where } s = \frac{a+b+c}{2}.$$

Solutions to Selected Exercises

5. Use the Law of Cosines to solve the triangle: $a = 9$, $b = 12$, $c = 15$.

   Solution:

   $\cos A = \dfrac{144 + 225 - 81}{360} = 0.8$

   $A \approx 36.9°$

   $\cos B = \dfrac{81 + 225 - 144}{270} = 0.6$

   $B \approx 53.1°$

   $C = 180° - (36.9° + 53.1°)$

   $C = 90°$

7. Use the Law of Cosines to solve the triangle: $a = 75.4$, $b = 52$, $c = 52$.

Solution:

$$\cos A = \frac{(52)^2 + (52)^2 - (75.4)^2}{2(52)(52)} = -0.05125$$

$$A \approx 92.9°$$

Since $b = c$, the triangle is isosceles and $B = C$.

$$2B = 180° - 92.9°$$

$$B = C = 43.55°$$

13. Use the Law of Cosines to solve the triangle: $C = 125°40'$, $a = 32$, $b = 32$.

Solution:

Since $a = b$, the triangle is isosceles and $A = B$.

$$2A = 180° - 125°40'$$

$$A = B = 27°10'$$

$$c^2 = (32)^2 + (32)^2 - 2(32)(32) \cos 125°40'$$

$$c \approx 56.9$$

17. Use Heron's Formula to find the area of the triangle: $a = 12$, $b = 15$, $c = 9$.

Solution:

$$s = \tfrac{1}{2}(12 + 15 + 9) = 18$$

$$\text{Area} = \sqrt{18(18 - 12)(18 - 15)(18 - 9)}$$

$$= \sqrt{2916}$$

$$= 54 \text{ square units}$$

23. Two ships leave a port at 9 A. M. One travels at a bearing of N 53° W at 12 miles per hour and the other at a bearing of S 67° W at 16 miles per hour. Approximately how far apart are they at noon of that day?

Solution:

By noon, the first ship has traveled $3(12) = 36$ miles, and the second ship has traveled $3(16) = 48$ miles.

$$d^2 = 36^2 + 48^2 - 2(36)(48) \cos 60°$$

$$d^2 = 1872$$

$$d \approx 43.3 \text{ mi}$$

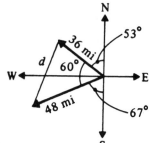

27. Determine the angle θ as shown on the streetlight in the figure.

Solution:

$$\cos \theta = \frac{3^2 + 2^2 - 4.25^2}{2(3)(2)} = -0.421875$$

$$\theta \approx 114.95°$$

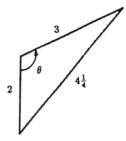

31. In a (square) baseball diamond with 90-foot sides the pitcher's mound is 60 feet from home plate.

(a) How far is it from the pitcher's mound to third base?
(b) When a runner is halfway from second to third, how far is the runner from the pitcher's mound?

Solution:

(a) $x^2 = 90^2 + 60^2 - 2(90)(60) \cos 45°$

     $x \approx 63.7$ feet

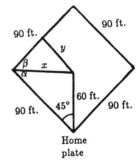

(b) $\dfrac{60}{\sin \alpha} = \dfrac{63.7}{\sin 45°}$

    $\sin \alpha = \dfrac{60 \sin 45°}{63.7}$

      $\alpha \approx 41.73°$

      $\beta = 90° - \alpha = 48.27°$

      $y^2 = 45^2 + 63.7^2 - 2(45)(63.7) \cos 48.27°$

      $y \approx 47.6$ feet

37. Use the Law of Cosines to prove that $\dfrac{1}{2}bc(1 + \cos A) = \dfrac{a+b+c}{2} \cdot \dfrac{-a+b+c}{2}$.

Solution:

$$\frac{1}{2}bc(1 + \cos A) = \frac{1}{2}bc\left[1 + \frac{b^2 + c^2 - a^2}{2bc}\right]$$

$$= \frac{1}{2}bc\left[\frac{2bc + b^2 + c^2 - a^2}{2bc}\right]$$

$$= \frac{1}{4}[(b+c)^2 - a^2]$$

$$= \frac{1}{4}[(b+c) + a][(b+c) - a]$$

$$= \frac{b+c+a}{2} \cdot \frac{b+c-a}{2}$$

$$= \frac{a+b+c}{2} \cdot \frac{-a+b+c}{2}$$

SECTION 4.3

Vectors

- A vector $\mathbf{v}$ is the collection of all directed line segments that are equivalent to a given directed line segment $\overrightarrow{PQ}$.

- You should be able to *geometrically* perform the operations of vector addition and scalar multiplication.

- The component form of the vector with initial point $P = (p_1,\ p_2)$ and terminal point $Q = (q_1,\ q_2)$ is

$$\overrightarrow{PQ} = \langle q_1 - p_1,\ q_2 - p_2 \rangle = \langle v_1,\ v_2 \rangle = \mathbf{v}.$$

- The magnitude of $\mathbf{v} = \langle v_1,\ v_2 \rangle$ is given by $\|\mathbf{v}\| = \sqrt{v_1{}^2 + v_2{}^2}$.

- You should be able to perform the operations of scalar multiplication and vector addition in component form.

- You should know the following properties of vector addition and scalar multiplication.
  (a) $\mathbf{u} + \mathbf{v} = \mathbf{v} + \mathbf{u}$
  (b) $(\mathbf{u} + \mathbf{v}) + \mathbf{w} = \mathbf{u} + (\mathbf{v} + \mathbf{w})$
  (c) $\mathbf{u} + \mathbf{O} = \mathbf{u}$
  (d) $\mathbf{u} + (-\mathbf{u}) = \mathbf{O}$
  (e) $c(d\mathbf{u}) = (cd)\mathbf{u}$
  (f) $(c + d)\mathbf{u} = c\mathbf{u} + d\mathbf{u}$
  (g) $c(\mathbf{u} + \mathbf{v}) = c\mathbf{u} + c\mathbf{v}$
  (h) $1(\mathbf{u}) = \mathbf{u},\ 0\mathbf{u} = \mathbf{O}$
  (i) $\|c\mathbf{v}\| = |c|\,\|\mathbf{v}\|$

- A unit vector in the direction of $\mathbf{v}$ is given by $\mathbf{u} = \dfrac{\mathbf{v}}{\|\mathbf{v}\|}$.

- The standard unit vectors are $\mathbf{i} = \langle 1,\ 0 \rangle$ and $\mathbf{j} = \langle 0,\ 1 \rangle$. $\mathbf{v} = \langle v_1,\ v_2 \rangle$ can be written as $\mathbf{v} = v_1\mathbf{i} + v_2\mathbf{j}$.

- A vector $\mathbf{v}$ with magnitude $\|\mathbf{v}\|$ and direction θ can be written as $\mathbf{v} = a\mathbf{i} + b\mathbf{j} = \|\mathbf{v}\|(\cos\theta)\mathbf{i} + \|\mathbf{v}\|(\sin\theta)\mathbf{j}$ where $\tan\theta = b/a$.

Solutions to Selected Exercises

3. Use the figure to sketch a graph of $u + v$.

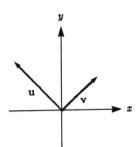

Solution:
Move the initial point of v to the terminal point of u.

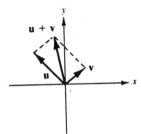

9. Find the component form and the magnitude of the vector v.

Solution:

$$v = \langle -1 - 2,\ 3 - 1 \rangle = \langle -3,\ 2 \rangle$$
$$\|v\| = \sqrt{(-3)^2 + (2)^2} = \sqrt{13}$$

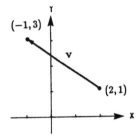

13. Find the component form and the magnitude of the vector v with initial point $(-1,\ 5)$ and terminal point $(15,\ 2)$.

Solution:

$$v = \langle 15 - (-1),\ 2 - 5 \rangle = \langle 16,\ -3 \rangle$$
$$\|v\| = \sqrt{(16)^2 + (-3)^2} = \sqrt{265}$$

19. Find (a) $u + v$, (b) $u - v$, and (c) $2u - 3v$ for $u = \langle -2,\ 3 \rangle$, $v = \langle -2,\ 1 \rangle$.

Solution:
(a) $u + v = \langle -2 + (-2),\ 3 + 1 \rangle = \langle -4,\ 4 \rangle$
(b) $u - v = \langle -2 - (-2),\ 3 - 1 \rangle = \langle 0,\ 2 \rangle$
(c) $2u - 3v = \langle 2(-2) - 3(-2),\ 2(3) - 3(1) \rangle = \langle 2,\ 3 \rangle$

25. Find (a) $\mathbf{u}+\mathbf{v}$, (b) $\mathbf{u}-\mathbf{v}$, and (c) $2\mathbf{u}-3\mathbf{v}$ for $\mathbf{u}=2\mathbf{i}$, $\mathbf{v}=\mathbf{j}$.

Solution:

(a) $\mathbf{u}+\mathbf{v}=2\mathbf{i}+\mathbf{j}$
(b) $\mathbf{u}-\mathbf{v}=2\mathbf{i}-\mathbf{j}$
(c) $2\mathbf{u}-3\mathbf{v}=4\mathbf{i}-3\mathbf{j}$

29. Sketch $\mathbf{v}$ and find its component form given $\|\mathbf{v}\|=1$, $\theta=150°$. (Assume θ is measured counterclockwise from the x-axis to the vector.)

Solution:

$$\mathbf{v}=1\cos 150°\mathbf{i}+1\sin 150°\mathbf{j}$$

$$=-\frac{\sqrt{3}}{2}\mathbf{i}+\frac{1}{2}\mathbf{j}$$

$$=\left\langle-\frac{\sqrt{3}}{2},\ \frac{1}{2}\right\rangle$$

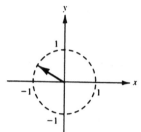

33. Sketch $\mathbf{v}$ and find its component form given $\|\mathbf{v}\|=2$, and $\mathbf{v}$ is in the direction of $\mathbf{i}+3\mathbf{j}$. (Assume θ is measured counterclockwise from the x-axis to the vector.)

Solution:

$$\tan\theta=\frac{3}{1}\Rightarrow\sin\theta=\frac{3\sqrt{10}}{10}\text{ and }\cos\theta=\frac{\sqrt{10}}{10}$$

$$\mathbf{v}=2\left(\frac{\sqrt{10}}{10}\right)\mathbf{i}+2\left(\frac{3\sqrt{10}}{10}\right)\mathbf{j}$$

$$=\left\langle\frac{\sqrt{10}}{5},\ \frac{3\sqrt{10}}{5}\right\rangle$$

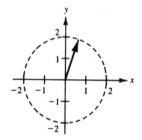

37. Find the component form of $\mathbf{v}=\mathbf{u}+2\mathbf{w}$ and sketch the indicated vector operations geometrically, where $\mathbf{u}=2\mathbf{i}-\mathbf{j}$ and $\mathbf{w}=\mathbf{i}+2\mathbf{j}$.

Solution:

$$\mathbf{v}=\mathbf{u}+2\mathbf{w}$$

$$=4\mathbf{i}+3\mathbf{j}$$

$$=\langle 4,\ 3\rangle$$

41. Find the component form of the sum of the vectors $\mathbf{u}$ and $\mathbf{v}$ with direction angles $\theta_{\mathbf{u}}$ and $\theta_{\mathbf{v}}$, respectively, given $\|\mathbf{u}\|=5$, $\theta_{\mathbf{u}}=0°$, and $\|\mathbf{v}\|=5$, $\theta_{\mathbf{v}}=90°$.

Solution:

$$\|\mathbf{u}\|=5,\ \theta_{\mathbf{u}}=0°\Rightarrow\mathbf{u}=5\mathbf{i}$$

$$\|\mathbf{v}\|=5,\ \theta_{\mathbf{v}}=90°\Rightarrow\mathbf{v}=5\mathbf{j}$$

$$\mathbf{u}+\mathbf{v}=5\mathbf{i}+5\mathbf{j}=\langle 5,\ 5\rangle$$

45. Find a unit vector in the direction of $\mathbf{v} = 4\mathbf{i} - 3\mathbf{j}$.

Solution:

$$\|\mathbf{v}\| = \sqrt{(4)^2 + (-3)^2} = 5$$

$$\frac{\mathbf{v}}{\|\mathbf{v}\|} = \frac{4\mathbf{i} - 3\mathbf{j}}{5} = \left\langle \frac{4}{5}, -\frac{3}{5} \right\rangle$$

49. Use the Law of Cosines to find the angle α between the vectors $\mathbf{v} = \mathbf{i} + \mathbf{j}$ and $\mathbf{w} = 2(\mathbf{i} - \mathbf{j})$. (Assume $0° \le \alpha \le 180°$.)

Solution:

$\mathbf{v} = \mathbf{i} + \mathbf{j}, \quad \mathbf{w} = 2(\mathbf{i} - \mathbf{j})$

$\mathbf{u} = \mathbf{w} - \mathbf{v} = \mathbf{i} - 3\mathbf{j}$

$\|\mathbf{v}\| = \sqrt{2}$

$\|\mathbf{w}\| = 2\sqrt{2}$

$\|\mathbf{u}\| = \sqrt{10}$

$\cos\theta = \dfrac{2 + 8 - 10}{2(\sqrt{2})(2\sqrt{2})} = 0$

$\theta = 90°$

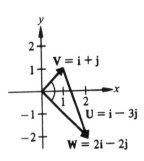

53. Two forces, one of 35 pounds and the other of 50 pounds, act on the same object. The angle between the forces is $30°$. Find the magnitude of the resultant (vector sum) of these forces.

Solution:

$\mathbf{u} = 35\mathbf{i}$

$\mathbf{v} = 50(\cos 30°\mathbf{i} + \sin 30°\mathbf{j}) = 50\left(\dfrac{\sqrt{3}}{2}\mathbf{i} + \dfrac{1}{2}\mathbf{j}\right)$

$\mathbf{r} = \mathbf{u} + \mathbf{v} = (35 + 25\sqrt{3})\mathbf{i} + 25\mathbf{j}$

$\|\mathbf{r}\| \approx 82.2 \text{ lb}$

57. A heavy implement is dragged 10 feet across the floor, using a force of 85 pounds. Find the work done if the direction of the force is $60°$ above the horizontal (see figure). (Use the formula for work, $W = FD$, where F is the horizontal component of force and D is the horizontal distance.)

Solution:
The horizontal component of the force is $85 \cos 60° = \frac{85}{2}$.

$$W = FD = \frac{85}{2}(10) = 425 \text{ ft-lb}$$

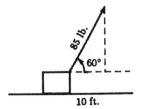

61. An airplane is flying in the direction S 32° E, with an airspeed of 540 miles per hour. Because of the wind, its groundspeed and direction are 500 miles per hour and S 40° E, respectively. Find the direction and speed of the wind.

Solution:

$$a = |\overrightarrow{BC}| = \text{speed of wind}$$

$$a^2 = 500^2 + 540^2 - 2(500)(540)\cos 8°$$

$$a \approx 82.8 \text{ mi/hr}$$

$$\cos C \approx \frac{82.8^2 + 500^2 - 540^2}{2(82.8)(500)}$$

$$C \approx 114.8°$$

$$C + \theta + 40 \approx 180°$$

$$\theta \approx \text{ N } 25.2° \text{ E}$$

SECTION 4.4

The Dot Product

- The *dot product* of $\mathbf{u} = \langle u_1, u_2 \rangle$ and $\mathbf{v} = \langle v_1, v_2 \rangle$ is

$$\mathbf{u} \cdot \mathbf{v} = u_1 v_1 + u_2 v_2.$$

- Know the following properties of the dot product.
  - (a) $\mathbf{u} \cdot \mathbf{v} = \mathbf{v} \cdot \mathbf{u}$
  - (b) $\mathbf{u} \cdot (\mathbf{v} + \mathbf{w}) = \mathbf{u} \cdot \mathbf{v} + \mathbf{u} \cdot \mathbf{w}$
  - (c) $c(\mathbf{u} \cdot \mathbf{v}) = (c\mathbf{u}) \cdot \mathbf{v} = \mathbf{u} \cdot (c\mathbf{v})$
  - (d) $\mathbf{O} \cdot \mathbf{v} = 0$
  - (e) $\mathbf{v} \cdot \mathbf{v} = \|\mathbf{v}\|^2$

- If θ is the angle between two nonzero vectors $\mathbf{u}$ and $\mathbf{v}$, then

$$\cos \theta = \frac{\mathbf{u} \cdot \mathbf{v}}{\|\mathbf{u}\| \, \|\mathbf{v}\|}.$$

- $\mathbf{u}$ and $\mathbf{v}$ are *orthogonal* if $\mathbf{u} \cdot \mathbf{v} = 0$.

- $\mathbf{u}$ and $\mathbf{v}$ are *parallel* if $\mathbf{u} = k\mathbf{v}$, $\quad k \neq 0$.

- The projection of $\mathbf{u}$ onto $\mathbf{v}$, where $\mathbf{u}$ and $\mathbf{v}$ are nonzero vectors, is

$$\text{proj}_{\mathbf{v}}(\mathbf{u}) = \left(\frac{\mathbf{u} \cdot \mathbf{v}}{\|\mathbf{v}\|^2} \right) \mathbf{v}.$$

- $k = \dfrac{\mathbf{u} \cdot \mathbf{v}}{\|\mathbf{v}\|} = \|\mathbf{u}\| \cos \theta$ is called the component of $\mathbf{u}$ in the direction of $\mathbf{v}$.

- To find work W:
  - (a) $W = \|\text{proj}_{\overrightarrow{PQ}}(\mathbf{F})\| \, \|\overrightarrow{PQ}\|$
  - (b) $W = \mathbf{F} \cdot \overrightarrow{PQ}$

Solutions to Selected Exercises

3. Find (a) $\mathbf{u} \cdot \mathbf{v}$, (b) $\mathbf{u} \cdot \mathbf{u}$, (c) $\|\mathbf{u}\|^2$, (d) $(\mathbf{u} \cdot \mathbf{v})\mathbf{v}$, and (e) $\mathbf{u} \cdot (2\mathbf{v})$ for $\mathbf{u} = \langle 2, -3 \rangle$ and $\mathbf{v} = \langle 0, 6 \rangle$.

Solution:

(a) $\mathbf{u} \cdot \mathbf{v} = 2(0) + (-3)(6) = -18$

(b) $\mathbf{u} \cdot \mathbf{u} = 2(2) + (-3)(-3) = 13$

(c) $\|\mathbf{u}\|^2 = (2)^2 + (-3)^2 = 13$

(d) $(\mathbf{u} \cdot \mathbf{v})\mathbf{v} = -18\mathbf{v} = \langle 0, -108 \rangle$

(e) $\mathbf{u} \cdot (2\mathbf{v}) = 2((2)(0)) + (-3)(2(6)) = -36$

9. Find the angle θ between the vectors $\mathbf{u} = 3\mathbf{i} + \mathbf{j}$ and $\mathbf{v} = -2\mathbf{i} + 4\mathbf{j}$.

Solution:

$$\cos\theta = \frac{\mathbf{u} \cdot \mathbf{v}}{\|\mathbf{u}\|\,\|\mathbf{v}\|}$$

$$= \frac{3(-2) + (1)(4)}{\sqrt{(3)^2 + (1)^2}\sqrt{(-2)^2 + (4)^2}}$$

$$= \frac{-2}{\sqrt{10}\sqrt{20}} = \frac{-2}{10\sqrt{2}} = -\frac{\sqrt{2}}{10}$$

$$\theta = \arccos\left(-\frac{\sqrt{2}}{10}\right) \approx 98.13°$$

13. Find the angle θ between the vectors $\mathbf{u} = \cos(\pi/4)\mathbf{i} + \sin(\pi/4)\mathbf{j}$ and $\mathbf{v} = \cos(\pi/2)\mathbf{i} + \sin(\pi/2)\mathbf{j}$.

Solution:

$$\mathbf{u} = \cos\left(\frac{\pi}{4}\right)\mathbf{i} + \sin\left(\frac{\pi}{4}\right)\mathbf{j} = \frac{\sqrt{2}}{2}\mathbf{i} + \frac{\sqrt{2}}{2}\mathbf{j}$$

$$\mathbf{v} = \cos\left(\frac{\pi}{2}\right)\mathbf{i} + \sin\left(\frac{\pi}{2}\right)\mathbf{j} = \mathbf{j}$$

$$\cos\theta = \frac{\mathbf{u} \cdot \mathbf{v}}{\|\mathbf{u}\|\,\|\mathbf{v}\|}$$

$$= \frac{(\sqrt{2}/2)(0) + (\sqrt{2}/2)(1)}{\sqrt{(\sqrt{2}/2)^2 + (\sqrt{2}/2)^2}\sqrt{(0)^2 + (1)^2}}$$

$$= \frac{\sqrt{2}/2}{\sqrt{1}\sqrt{1}} = \frac{\sqrt{2}}{2}$$

$$\theta = \arccos\frac{\sqrt{2}}{2} = 45°$$

15. Determine whether **u** and **v** are orthogonal, parallel, or neither.

$$\mathbf{u} = \langle 4,\ 0 \rangle, \quad \mathbf{v} = \langle 1,\ 1 \rangle$$

Solution:

$\mathbf{u} \cdot \mathbf{v} = 4 \neq 0$. Therefore, **u** and **v** are not orthogonal.

$k\mathbf{v} = \langle k,\ k \rangle \neq \mathbf{u}$ for any real number k. Therefore, **u** and **v** are not parallel.

Neither

21. Determine whether **u** and **v** are orthogonal, parallel, or neither.

$$\mathbf{u} = \langle 2,\ -2 \rangle, \quad \mathbf{v} = \langle -1,\ -1 \rangle$$

Solution:

$\mathbf{u} \cdot \mathbf{v} = 0$. Therefore, **u** and **v** are orthogonal.

23. (a) Find the projection of **u** onto **v**, and (b) the vector component of **u** orthogonal to **v**.

$$\mathbf{u} = \langle 2,\ 3 \rangle, \quad \mathbf{v} = \langle 5,\ 1 \rangle$$

Solution:

(a) $\text{proj}_{\mathbf{v}}(\mathbf{u}) = \left(\dfrac{\mathbf{u} \cdot \mathbf{v}}{\|\mathbf{v}\|^2} \right)\mathbf{v} = \dfrac{13}{26}\mathbf{v} = \dfrac{1}{2}\langle 5,\ 1 \rangle = \left\langle \dfrac{5}{2},\ \dfrac{1}{2} \right\rangle$

(b) $\mathbf{u} - \text{proj}_{\mathbf{v}}(\mathbf{u}) = \langle 2,\ 3 \rangle - \left\langle \dfrac{5}{2},\ \dfrac{1}{2} \right\rangle = \left\langle -\dfrac{1}{2},\ \dfrac{5}{2} \right\rangle$

27. (a) Find the projection of **u** onto **v**, and (b) the vector component of **u** orthogonal to **v**.

$$\mathbf{u} = \langle 1,\ 1 \rangle, \quad \mathbf{v} = \langle -2,\ -1 \rangle$$

Solution:

(a) $\text{proj}_{\mathbf{v}}(\mathbf{u}) = \left(\dfrac{\mathbf{u} \cdot \mathbf{v}}{\|\mathbf{v}\|^2} \right)\mathbf{v} = \dfrac{-3}{5}\mathbf{v} = \left\langle \dfrac{6}{5},\ \dfrac{3}{5} \right\rangle$

(b) $\mathbf{u} - \text{proj}_{\mathbf{v}}(\mathbf{u}) = \langle 1,\ 1 \rangle - \left\langle \dfrac{6}{5},\ \dfrac{3}{5} \right\rangle = \left\langle -\dfrac{1}{5},\ \dfrac{2}{5} \right\rangle$

33. An implement is dragged 10 feet across a floor, using a force of 85 pounds. Find the work done if the direction of the force is 60° above the horizontal (see figure).

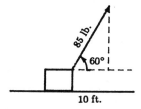

10 ft.

Solution:

$$W = \mathbf{F} \cdot \overrightarrow{PQ}$$
$$= (\cos \theta) \|\mathbf{F}\| \|\overrightarrow{PQ}\|$$
$$= (\cos 60°)(85)(10)$$
$$= \left(\frac{1}{2}\right)(85)(10)$$
$$= 425 \text{ ft-lb}$$

35. Find the work done in moving a particle from P to Q if the magnitude and direction of the force is given by $\mathbf{v}$.

$$P = (0,\ 0), \quad Q = (4,\ 7), \quad \mathbf{v} = \langle 1,\ 4 \rangle$$

Solution:

$$\overrightarrow{PQ} = \langle 4,\ 7 \rangle$$
$$W = \mathbf{v} \cdot \overrightarrow{PQ} = 4 + 28 = 32$$

REVIEW EXERCISES FOR CHAPTER 4

Solutions to Selected Exercises

5. Solve the triangle, given $B = 110°$, $a = 4$, $c = 4$.

Solution:

Since the triangle is isosceles,

$$A = C = \frac{1}{2}(180 - 110) = 35°.$$

By the Law of Sines:

$$\frac{4}{\sin 35°} = \frac{b}{\sin 110°}$$
$$b \approx 6.6$$

9. Solve the triangle, given $B = 115°$, $a = 7$, $b = 14.5$.

Solution:

By the Law of Sines:

$$\frac{\sin A}{7} = \frac{\sin 115°}{14.5}$$
$$\sin A \approx 0.4375$$
$$A \approx 25.9°$$
$$C = 180 - (115 + 25.9)$$
$$C = 39.1°$$
$$\frac{c}{\sin 39.1°} = \frac{14.5}{\sin 115°}$$
$$c \approx 10.1$$

13. Solve the triangle, given $B = 150°$, $a = 10$, $c = 20$.

Solution:

By the Law of Cosines:

$$b^2 = 10^2 + 20^2 - 2(10)(20) \cos 150°$$
$$b \approx 29.1$$

By the Law of Sines:

$$\frac{\sin A}{10} = \frac{\sin 150°}{29.1}$$
$$\sin A \approx 0.1719$$
$$A \approx 9.9°$$
$$C = 180 - (150 + 9.9)$$
$$C = 20.1°$$

17. Find the area of the triangle with $a = 4$, $b = 5$, and $c = 7$.

Solution:

$$s = \frac{4 + 5 + 7}{2} = 8$$
$$A = \sqrt{8(8 - 4)(8 - 5)(8 - 7)} = \sqrt{96} = 4\sqrt{6} \text{ square units}$$

21. Find the height of a tree that stands on a hillside of slope 32° (from the horizontal) if from a point 75 feet downhill the angle of elevation to the top of the tree is 48° (see figure).

Solution:

Let $x =$ the height of the hillside.

$$\sin 32° = \frac{x}{75}$$
$$x = 75 \sin 32 \approx 39.7439 \text{ feet}$$

Let $y =$ the horizontal distance.

$$y = \sqrt{75^2 - x^2} = \sqrt{75^2 - 39.7439^2} \approx 63.6036 \text{ feet}$$

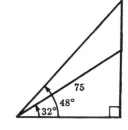

Let $h =$ the height of the tree.

$$\tan 48° = \frac{x + h}{y}$$
$$\tan 48° = \frac{39.7439 + h}{63.6036}$$
$$h = 63.6036 \tan 48° - 39.7439$$
$$h \approx 31 \text{ feet}$$

23. From a certain distance, the angle of elevation of the top of a building is 17°. At a point 50 meters closer to the building, the angle of elevation is 31°. Approximate the height of the building.

Solution:

$$\tan 17° = \frac{h}{50 + y} \Rightarrow h = (50 + y)\tan 17°$$

$$\tan 31° = \frac{h}{y} \Rightarrow h = y\tan 31°$$

$$(50 + y)\tan 17° = y\tan 31°$$

$$50\tan 17° + y\tan 17° = y\tan 31°$$

$$y(\tan 17° - \tan 31°) = -50\tan 17°$$

$$y = \frac{-50\tan 17°}{\tan 17° - \tan 31°} \approx 51.7959 \text{ m}$$

$$h = y\tan 31° = 51.7959\tan 31° \approx 31.1 \text{ m}$$

27. Find the component form of the vector **v** with initial point $(0, 10)$, and terminal point $(7, 3)$.

Solution:

$$\mathbf{v} = \langle 7 - 0,\ 3 - 10 \rangle = \langle 7,\ -7 \rangle$$

33. Find the component form of $4\mathbf{u} - 5\mathbf{v}$ and sketch its graph given that $\mathbf{u} = 6\mathbf{i} - 5\mathbf{j}$ and $\mathbf{v} = 10\mathbf{i} + 3\mathbf{j}$.

Solution:

$$4\mathbf{u} - 5\mathbf{v} = (24\mathbf{i} - 20\mathbf{j}) - (50\mathbf{i} + 15\mathbf{j})$$

$$= -26\mathbf{i} - 35\mathbf{j}$$

$$= \langle -26,\ -35 \rangle$$

37. Find the direction and magnitude of the resultant of the three forces shown in the figure.

Solution:

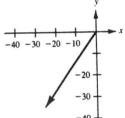

$$\|\mathbf{u}\| = 13$$

$$\|\mathbf{v}\| = 5$$

$$\mathbf{u} = 300\left(\frac{5}{13}\mathbf{i} + \frac{12}{13}\mathbf{j}\right)$$

$$\mathbf{v} = 150\left(-\frac{4}{5}\mathbf{i} + \frac{3}{5}\mathbf{j}\right)$$

$$\mathbf{w} = 250(0\mathbf{i} - \mathbf{j})$$

$$\tan \beta = \tfrac{3}{4} \qquad \tan \alpha = \tfrac{12}{5}$$

$$\mathbf{r} = \mathbf{u} + \mathbf{v} + \mathbf{w} = \left(\frac{1500}{13} - 120 + 0\right)\mathbf{i} + \left(\frac{3600}{13} + 90 - 250\right)\mathbf{j} = \frac{-60}{13}\mathbf{i} + \frac{1520}{13}\mathbf{j}$$

$$\|\mathbf{r}\| = \sqrt{\left(\frac{-60}{13}\right)^2 + \left(\frac{1520}{13}\right)^2} \approx 117.0 \text{ lb}$$

$$\theta = 180° - \arctan\frac{1520}{60} \approx 92.3°$$

41. Find the vector projection $\mathbf{w}$ of $\mathbf{u} = 3\mathbf{i} + 2\mathbf{j}$ onto $\mathbf{v} = \frac{1}{2}\mathbf{i} - \frac{3}{2}\mathbf{j}$.

Solution:

$$\mathbf{w} = \text{proj}_{\mathbf{v}}(\mathbf{u}) = \left(\frac{\mathbf{u} \cdot \mathbf{v}}{\|\mathbf{v}\|^2}\right)\mathbf{v}$$

$$= \frac{3(\frac{1}{2}) + 2(-\frac{3}{2})}{(\frac{1}{2})^2 + (-\frac{3}{2})^2}\mathbf{v}$$

$$= -\frac{3}{5}\mathbf{v}$$

$$= -\frac{3}{5}\left(\frac{1}{2}\mathbf{i} - \frac{3}{2}\mathbf{j}\right)$$

$$= -\frac{3}{10}(\mathbf{i} - 3\mathbf{j})$$

47. Use the result of Exercise 46 to reflect the vector $\mathbf{u} = \mathbf{i} + 3\mathbf{j}$ through the following vectors.

(a) $\mathbf{v} = \mathbf{i}$ (b) $\mathbf{v} = \mathbf{j}$ (c) $\mathbf{v} = \mathbf{i} + \mathbf{j}$

Solution:

$$\mathbf{w} = \left(\frac{2\mathbf{u} \cdot \mathbf{v}}{\mathbf{v} \cdot \mathbf{v}}\right)\mathbf{v} - \mathbf{u}$$

(a) $\mathbf{w} = 2\mathbf{v} - \mathbf{u} = \mathbf{i} - 3\mathbf{j}$
(b) $\mathbf{w} = 6\mathbf{v} - \mathbf{u} = -\mathbf{i} + 3\mathbf{j}$
(c) $\mathbf{w} = \frac{8}{2}\mathbf{v} - \mathbf{u} = 3\mathbf{i} + \mathbf{j}$

Practice Test for Chapter 4

For Exercises 1 and 2, use the Law of Sines to find the remaining sides and angles of the triangle.

1. $A = 40°$, $B = 12°$, $b = 100$

2. $C = 150°$, $a = 5$, $c = 20$

3. Find the area of the triangle: $a = 3$, $b = 5$, $C = 130°$.

4. Determine the number of solutions to the triangle: $a = 10$, $b = 35$, $A = 22.5°$.

For Exercises 5 and 6, use the Law of Cosines to find the remaining sides and angles of the triangle.

5. $a = 49$, $b = 53$, $c = 38$

6. $C = 29°$, $a = 100$, $b = 300$

7. Use Heron's Formula to find the area of the triangle: $a = 4.1$, $b = 6.8$, $c = 5.5$.

8. A ship travels 40 miles due east, then adjusts its course 12° southward. After traveling 70 miles in that direction, how far is the ship from its point of departure?

9. **w** is $4\mathbf{u} - 7\mathbf{v}$ where $\mathbf{u} = 3\mathbf{i} + \mathbf{j}$ and $\mathbf{v} = -\mathbf{i} + 2\mathbf{j}$. Find **w**.

10. Find a unit vector in the direction of $\mathbf{v} = 5\mathbf{i} - 3\mathbf{j}$.

11. Find the angle between $\mathbf{u} = 6\mathbf{i} + 5\mathbf{j}$ and $\mathbf{v} = 2\mathbf{i} - 3\mathbf{j}$.

12. **v** is a vector of magnitude 4 making an angle of 30° with the positive x-axis. Find **v**.

For Exercises 13–19, use the vectors $\mathbf{u} = \langle 3, -5 \rangle$ and $\mathbf{v} = \langle -2, 1 \rangle$.

13. Find $\mathbf{u} \cdot \mathbf{v}$.

14. Find $\|\mathbf{u}\|$.

15. Find the angle θ between **u** and **v**.

16. Find $(\mathbf{v} \cdot \mathbf{u})\mathbf{u}$.

17. Determine if **u** and **v** are orthogonal, parallel, or neither.

18. Find the projection of **u** onto **v**.

19. Find the vector component of **u** orthogonal to **v**.

20. An implement is dragged 20 feet across a floor, using a force of 70 pounds. Find the work done if the direction of the force is 40° above the horizontal.

CHAPTER 5

Complex Numbers

SECTION 5.1

Complex Numbers

■ You should know how to work with complex numbers.

■ Operations on Complex Numbers

(a) Addition: $(a + bi) + (c + di) = (a + c) + (b + d)i$

(b) Subtraction: $(a + bi) - (c + di) = (a - c) + (b - d)i$

(c) Multiplication: $(a + bi)(c + di) = (ac - bd) + (ad + bc)i$

(d) Division: $\dfrac{a + bi}{c + di} = \dfrac{a + bi}{c + di} \cdot \dfrac{c - di}{c - di} = \dfrac{ac + bd}{c^2 + d^2} + \dfrac{bc - ad}{c^2 + d^2}i$

■ The complex conjugate of $a + bi$ is $a - bi$:

$$(a + bi)(a - bi) = a^2 + b^2$$

■ The additive inverse of $a + bi$ is $-a - bi$.

■ The multiplicative inverse of $a + bi$ is

$$\dfrac{a - bi}{a^2 + b^2}.$$

■ $\sqrt{-a} = \sqrt{a}\,i$ for $a > 0$.

Solutions to Selected Exercises

1. Write out the first 16 positive powers of i and express each as i, $-i$, 1, or -1.

Solution:

$$
\begin{array}{lllll}
i = i & i^5 = i & i^9 = i & i^{13} = i & i^{17} = i \\
i^2 = -1 & i^6 = -1 & i^{10} = -1 & i^{14} = -1 & i^{18} = -1 \\
i^3 = -i & i^7 = -i & i^{11} = -i & i^{15} = -i & i^{19} = -i \\
i^4 = 1 & i^8 = 1 & i^{12} = 1 & i^{16} = 1 & i^{20} = 1
\end{array}
$$

5. Find real numbers a and b so that the equation $(a - 1) + (b + 3)i = 5 + 8i$ is true.

Solution:

$$(a-1) + (b+3)i = 5 + 8i$$

$$a - 1 = 5 \quad \Rightarrow \quad a = 6$$
$$b + 3 = 8 \quad \Rightarrow \quad b = 5$$

9. Write $2 - \sqrt{-27}$ in standard form and find its complex conjugate.

Solution:

$$2 - \sqrt{-27} = 2 - \sqrt{27}\,i = 2 - 3\sqrt{3}\,i$$

Complex conjugate: $2 + 3\sqrt{3}\,i$

13. Write $-6i + i^2$ in standard form and find its complex conjugate.

Solution:

$$-6i + i^2 = -6i + (-1) = -1 - 6i$$

Complex conjugate: $-1 + 6i$

21. Perform the indicated operation and write the result in standard form.

$$(8 - i) - (4 - i)$$

Solution:

$$(8 - i) - (4 - i) = 8 - i - 4 + i = 4$$

23. Perform the indicated operation and write the result in standard form.

$$(-2 + \sqrt{-8}) + (5 - \sqrt{-50})$$

Solution:

$$(-2 + \sqrt{-8}) + (5 - \sqrt{-50}) = -2 + 2\sqrt{2}\,i + 5 - 5\sqrt{2}\,i = 3 - 3\sqrt{2}\,i$$

27. Perform the indicated operation and write the result in standard form.

$$\sqrt{-6}\sqrt{-2}$$

Solution:

$$\sqrt{-6}\sqrt{-2} = (\sqrt{6}\,i)(\sqrt{2}\,i) = \sqrt{12}\,i^2 = 2\sqrt{3}(-1) = -2\sqrt{3}$$

31. Perform the indicated operation and write the result in standard form.

$$(1+i)(3-2i)$$

Solution:

$$(1+i)(3-2i) = 3 - 2i + 3i - 2i^2 = 3 + i + 2 = 5 + i$$

35. Perform the indicated operation and write the result in standard form.

$$6i(5-2i)$$

Solution:

$$6i(5-2i) = 30i - 12i^2 = 12 + 30i$$

39. Perform the indicated operation and write the result in standard form.

$$(\sqrt{14} + \sqrt{10}\,i)(\sqrt{14} - \sqrt{10}\,i)$$

Solution:

$$(\sqrt{14} + \sqrt{10}\,i)(\sqrt{14} - \sqrt{10}\,i) = 14 - 10i^2 = 14 + 10 = 24$$

45. Perform the indicated operation and write the result in standard form.

$$\frac{2+i}{2-i}$$

Solution:

$$\frac{2+i}{2-i} = \frac{2+i}{2-i} \cdot \frac{2+i}{2+i} = \frac{4 + 4i + i^2}{4+1} = \frac{3+4i}{5} = \frac{3}{5} + \frac{4}{5}i$$

53. Perform the indicated operation and write the result in standard form.

$$\frac{(21-7i)(4+3i)}{2-5i}$$

Solution:

$$\frac{(21-7i)(4+3i)}{2-5i} = \frac{(84 + 63i - 28i - 21i^2)}{2-5i} \cdot \frac{2+5i}{2+5i}$$

$$= \frac{(105 + 35i)(2+5i)}{4+25}$$

$$= \frac{210 + 525i + 70i + 175i^2}{29}$$

$$= \frac{35 + 595i}{29} = \frac{35}{29} + \frac{595}{29}i$$

59. Use the quadratic formula to solve $4x^2 + 16x + 17 = 0$.

Solution:

$4x^2 + 16x + 17 = 0$; $a = 4$, $b = 16$, $c = 17$

$$x = \frac{-16 \pm \sqrt{(16)^2 - 4(4)(17)}}{2(4)}$$

$$= \frac{-16 \pm \sqrt{-16}}{8} = \frac{-16 \pm 4i}{8}$$

$$= -2 \pm \frac{1}{2}i$$

63. Use the quadratic formula to solve $16t^2 - 4t + 3 = 0$.

Solution:

$16t^2 - 4t + 3 = 0$; $a = 16$, $b = -4$, $c = 3$

$$x = \frac{-(-4) \pm \sqrt{(-4)^2 - 4(16)(3)}}{2(16)}$$

$$= \frac{4 \pm \sqrt{-176}}{32} = \frac{4 \pm 4\sqrt{11}i}{32}$$

$$= \frac{1}{8} \pm \frac{\sqrt{11}}{8}i$$

65. Prove that the sum of a complex number and its conjugate is a real number.

Solution:

$$(a + bi) + (a - bi) = (a + a) + (b - b)i$$
$$= 2a + 0i = 2a \quad \text{which is a real number.}$$

69. Prove that the conjugate of the sum of two complex numbers is the sum of their conjugates.

Solution:

$$(a + bi) + (c + di) = (a + c) + (b + d)i$$

The complex conjugate of the sum is $(a + c) - (b + d)i$, and the sum of the conjugates is

$$(a - bi) + (c - di) = (a + c) + (-b - d)i$$
$$= (a + c) - (b + d)i$$

Thus, the conjugate of the sum is the sum of the conjugates.

SECTION 5.2

Complex Solutions of Equations

- A polynomial of degree n has exactly n solutions in the complex number system. These solutions may be real or complex and may be repeated.

- If $a + bi$, $(b \neq 0)$ is a zero of a polynomial function with real coefficients, then so is its conjugate, $a - bi$.

Solutions to Selected Exercises

3. Use the discriminant to determine the number of real solutions of $3x^2 + 4x + 1 = 0$.

Solution:

$a = 3,\ b = 4,\ c = 1$

$b^2 - 4ac = (4)^2 - 4(3)(1) = 4 > 0$

There are *two* real solutions.

7. Use the discriminant to determine the number of real solutions of $\frac{1}{5}x^2 + \frac{6}{5}x - 8 = 0$.

Solution:

$a = \frac{1}{5},\ b = \frac{6}{5},\ c = -8$

$b^2 - 4ac = \left(\frac{6}{5}\right)^2 - 4\left(\frac{1}{5}\right)(-8) = \frac{36}{25} + \frac{32}{5} = \frac{196}{25} > 0$

There are *two* real solutions.

9. Solve the quadratic equation $x^2 - 5 = 0$. List any complex solutions in $a + bi$ form.

Solution:

$$x^2 - 5 = 0$$
$$x^2 = 5$$
$$x = \pm\sqrt{5}$$

13. Solve the quadratic equation $x^2 - 8x + 16 = 0$. List any complex solutions in $a + bi$ form.

Solution:

$$x^2 - 8x + 16 = 0$$
$$(x - 4)^2 = 0$$
$$x = 4$$

19. Solve the quadratic equation $230 + 20x - 0.5x^2 = 0$. List any complex solutions in $a + bi$ form.

Solution:

$$230 + 20x - 0.5x^2 = 0$$
$$-0.5x^2 + 20x + 230 = 0$$

Multiply by -2 and complete the square.

$$x^2 - 40x - 460 = 0$$
$$x^2 - 40x = 460$$
$$x^2 - 40x + 400 = 460 + 400$$
$$(x - 20)^2 = 860$$
$$x - 20 = \pm\sqrt{860}$$
$$x - 20 + \pm 2\sqrt{215}$$
$$x = 20 \pm 2\sqrt{215}$$

21. Find a polynomial with integer coefficients that has the zeros 1, $5i$, and $-5i$.

Solution:

$$f(x) = (x - 1)(x - 5i)(x + 5i)$$
$$= (x - 1)(x^2 + 25)$$
$$= x^3 - x^2 + 25x - 25$$

25. Find a polynomial with integer coefficients that has the zeros i, $-i$, $6i$, and $-6i$.

Solution:

$$f(x) = (x - i)(x + i)(x - 6i)(x + 6i)$$
$$= (x^2 + 1)(x^2 + 36)$$
$$= x^4 + 37x^2 + 36$$

29. Find a polynomial with integer coefficients that has the zeros $\frac{3}{4}$, -2, and $-\frac{1}{2} + i$.

Solution:
Since $-\frac{1}{2} + i$ is a zero, so is $-\frac{1}{2} - i$.

$$f(x) = 16\left(x - \tfrac{3}{4}\right)(x + 2)\left[x - \left(-\tfrac{1}{2} + i\right)\right]\left[x - \left(-\tfrac{1}{2} - i\right)\right]$$
$$= 4(4x - 3)(x + 2)\left[x^2 + x + \left(\tfrac{1}{4} + 1\right)\right]$$
$$= (4x^2 + 5x - 6)(4x^2 + 4x + 5)$$
$$= 16x^4 + 36x^3 + 16x^2 + x - 30$$

33. Find all the zeros of $h(x) = x^2 - 4x + 1$ and write the polynomial as a product of linear factors.

Solution:

f has no rational zeros.

By the Quadratic Formula, the zeros are $x = \dfrac{4 \pm \sqrt{16-4}}{2} = 2 \pm \sqrt{3}$.

$$f(x) = [x - (2 + \sqrt{3})][x - (2 - \sqrt{3})] = (x - 2 - \sqrt{3})(x - 2 + \sqrt{3})$$

37. Find all the zeros of $f(z) = z^2 - 2z + 2$ and write the polynomial as a product of linear factors.

Solution:

f has no rational zeros.

By the Quadratic Formula, the zeros are $z = \dfrac{2 \pm \sqrt{4-8}}{2} = 1 \pm i$.

$$f(z) = [z - (1 + i)][z - (1 - i)] = (z - 1 - i)(z - 1 + i)$$

39. Find all the zeros of $g(x) = x^3 - 6x^2 + 13x - 10$ and write the polynomial as a product of linear factors.

Solution:

Possible rational zeros: ± 1, ± 2, ± 5, ± 10

Since $g(2) = 0$, $x - 2$ is a factor. By long division we have

$$
\begin{array}{r}
x^2 - 4x + 2 \\
x - 2 \overline{)\, x^3 - 6x^2 + 13x - 10} \\
\underline{x^3 - 2x^2} \\
-4x^2 + 13x \\
\underline{-4x^2 + 8x} \\
5x - 10 \\
\underline{5x - 10} \\
0
\end{array}
$$

Thus, $g(x) = (x - 2)(x^2 - 4x + 5)$ and by the Quadratic Formula, $x = \dfrac{4 \pm \sqrt{16 - 20}}{2} = 2 \pm i$ are also zeros.

$$g(x) = (x - 2)[x - (2 + i)][x - (2 - i)] = (x - 2)(x - 2 - i)(x - 2 + i)$$

45. Find all the zeros of $f(x) = 16x^3 - 20x^2 - 4x + 15$ and write the polynomial as a product of linear factors.

Solution:

Possible rational zeros: ± 1, ± 3, ± 5, ± 15, $\pm\dfrac{1}{2}$, $\pm\dfrac{3}{2}$, $\pm\dfrac{5}{2}$, $\pm\dfrac{15}{2}$, $\pm\dfrac{1}{4}$, $\pm\dfrac{3}{4}$, $\pm\dfrac{5}{4}$, $\pm\dfrac{15}{4}$, $\pm\dfrac{1}{8}$, $\pm\dfrac{3}{8}$,

$$\pm\dfrac{5}{8}, \ \pm\dfrac{15}{8}, \ \pm\dfrac{1}{16}, \ \pm\dfrac{3}{16}, \ \pm\dfrac{5}{16}, \ \pm\dfrac{15}{16}$$

Since $f\left(-\frac{3}{4}\right) = 0$, $\ \ 4x + 3$ is a factor. By long division we have

$$
\begin{array}{r}
4x^2 - 8x + 5 \\
4x + 3 \ \overline{)\ 16x^3 - 20x^2 - 4x + 15} \\
\underline{16x^3 + 12x^2} \\
-32x^2 - 4x \\
\underline{-32x^2 - 24x } \\
20x + 15 \\
\underline{20x + 15} \\
0
\end{array}
$$

Thus, $f(x) = (4x + 3)(4x^2 - 8x + 5)$ and by the Quadratic Formula, $x = \dfrac{8 \pm \sqrt{64 - 80}}{8} = 1 \pm \dfrac{1}{2}i$ are also zeros.

$$f(x) = (4x + 3)\left[x - \left(1 + \frac{1}{2}i\right)\right]\left[x - \left(1 - \frac{1}{2}i\right)\right] = (4x + 3)(2x - 2 - i)(2x - 2 + i)$$

51. Find all the zeros of $g(x) = x^4 - 4x^3 + 8x^2 - 16x + 16$ and write the polynomial as a product of linear factors.

Solution:

Possible rational zeros: ± 1, ± 2, ± 4, ± 8, ± 16

Since $g(2) = 0$, $\ \ x - 2$ is a factor. By long division we have

$$
\begin{array}{r}
x^3 - 2x^2 + 4x - 8 \\
x - 2 \ \overline{)\ x^4 - 4x^3 + 8x^2 - 16x + 16} \\
\underline{x^4 - 2x^3 } \\
-2x^3 + 8x^2 \\
\underline{-2x^3 + 4x^2 } \\
4x^2 - 16x \\
\underline{4x^2 - 8x } \\
-8x + 16 \\
\underline{-8x + 16} \\
0
\end{array}
$$

Thus, $g(x) = (x - 2)(x^3 - 2x^2 + 4x - 8)$. Now factoring by grouping, we have

$$g(x) = (x - 2)[x^2(x - 2) + 4(x - 2)]$$
$$= (x - 2)(x - 2)(x^2 + 4)$$
$$= (x - 2)^2(x + 2i)(x - 2i)$$

The zeros of g are 2 and $\pm 2i$.

59. Write $f(x) = x^4 - 4x^3 + 5x^2 - 2x - 6$
   (a) as the product of factors that are irreducible over the rationals,
   (b) as the product of linear and quadratic factors that are irreducible over the reals, and
   (c) in completely factored form. [*Hint:* One factor is $x^2 - 2x - 2$.]

Solution:

$$
\begin{array}{r}
x^2 - 2x + 3 \\
x^2 - 2x - 2 \overline{\smash{\big)}\ x^4 - 4x^3 + 5x^2 - 2x - 6} \\
\underline{-(x^4 - 2x^3 - 2x^2)} \\
-2x^3 + 7x^2 - 2x \\
\underline{-(-2x^3 + 4x^2 + 4x)} \\
3x^2 - 6x - 6 \\
\underline{-(3x^2 - 6x - 6)} \\
0
\end{array}
$$

$$f(x) = (x^2 - 2x + 3)(x^2 - 2x - 2)$$

(a) $f(x) = (x^2 - 2x + 3)(x^2 - 2x - 2)$
(b) $f(x) = (x^2 - 2x + 3)(x - 1 + \sqrt{3})(x - 1 - \sqrt{3})$
(c) $f(x) = (x - 1 + \sqrt{2}\,i)(x - 1 - \sqrt{2}\,i)(x - 1 + \sqrt{3})(x - 1 - \sqrt{3})$

Note: Use the Quadratic Formula for (b) and (c).

63. Use the zero, $r = 2i$, to find all the zeros of $f(x) = 2x^4 - x^3 + 7x^2 - 4x - 4$.

Solution:
Since $2i$ is a zero of f, so is $-2i$. Thus, $(x - 2i)(x + 2i) = x^2 + 4$ is a factor of $f(x)$. By long division we have

$$
\begin{array}{r}
2x^2 - x - 1 \\
x^2 + 4 \overline{\smash{\big)}\ 2x^4 - x^3 + 7x^2 - 4x - 4} \\
\underline{2x^4 \quad\quad + 8x^2} \\
-x^3 - x^2 - 4x \\
\underline{-x^3 \quad\quad - 4x} \\
-x^2 \quad\quad - 4 \\
\underline{-x^2 \quad\quad - 4} \\
0
\end{array}
$$

Thus, $f(x) = (x^2 + 4)(2x^2 - x - 1) = (x - 2i)(x + 2i)(2x + 1)(x - 1)$.
The zeros of f are $\pm 2i$, $-\frac{1}{2}$, and 1.

67. Use the zero, $r = -3 + \sqrt{2}\,i$, to find all the zeros of $f(x) = x^4 + 3x^3 - 5x^2 - 21x + 22$.

Solution:
Since $-3 + \sqrt{2}\,i$ is a zero of f, so is $-3 - \sqrt{2}\,i$.
Thus, $[x - (-3 + \sqrt{2}\,i)][x - (-3 - \sqrt{2}\,i)] = [(x + 3) - \sqrt{2}\,i)][(x + 3) + \sqrt{2}\,i]$

$$= (x + 3)^2 - 2i^2$$
$$= x^2 + 6x + 11$$

is a factor of x. Now, by long division we have

$$
\begin{array}{r}
x^2 - 3x + 2 \\
x^2 + 6x + 11 \overline{\smash{)}\ x^4 + 3x^3 - 5x^2 - 21x + 22} \\
\underline{x^4 + 6x^3 + 11x^2} \\
-3x^3 - 16x^2 - 21x \\
\underline{-3x^3 - 18x^2 - 33x} \\
2x^2 + 12x + 22 \\
\underline{2x^2 + 12x + 22} \\
0
\end{array}
$$

Thus, $f(x) = [x - (-3 + \sqrt{2}\,i)][x - (-3 - \sqrt{2}\,i)](x^2 - 3x + 2)$
$$= [x - (-3 + \sqrt{2}\,i)][x - (-3 - \sqrt{2}\,i)](x - 1)(x - 2)$$

The zeros of f are $-3 + \sqrt{2}\,i$, $-3 - \sqrt{2}\,i$, 1, and 2.

71. Find a quadratic function f (with integer coefficients) that has $\pm\sqrt{b}\,i$ as zeros. Assume that b is a positive integer.

Solution:
$$f(x) = (x - \sqrt{b}\,i)(x + \sqrt{b}\,i) = x^2 + b$$

SECTION 5.3

Trigonometric Form of Complex Numbers

- You should be able to graphically represent complex numbers and know the following facts about them.

- The absolute value of the complex number $z = a + bi$ is $|z| = \sqrt{a^2 + b^2}$.

- The trigonometric form of the complex number $z = a + bi$ is $z = r(\cos\theta + i\sin\theta)$ where
  - (a) $a = r\cos\theta$
  - (b) $b = r\sin\theta$
  - (c) $r = \sqrt{a^2 + b^2}$
  - (d) $\tan\theta = b/a$

- Given $z_1 = r_1(\cos\theta_1 + i\sin\theta_1)$ and $z_2 = r_2(\cos\theta_2 + i\sin\theta_2)$:
  - (a) $z_1 z_2 = r_1 r_2[\cos(\theta_1 + \theta_2) + i\sin(\theta_1 + \theta_2)]$
  - (b) $\dfrac{z_1}{z_2} = \dfrac{r_1}{r_2}[\cos(\theta_1 - \theta_2) + i\sin(\theta_1 - \theta_2)], \quad z \neq 0$

Solutions to Selected Exercises

3. Express the complex number $-3 - 3i$ in trigonometric form.

Solution:

$$z = -3 - 3i$$
$$r = \sqrt{(-3)^2 + (-3)^2} = \sqrt{18} = 3\sqrt{2}$$
$$\tan\theta = \frac{-3}{-3} = 1, \ \theta \text{ is in Quadrant III}$$
$$\theta = 225° \text{ or } \frac{5\pi}{4}$$
$$z = 3\sqrt{2}\left(\cos\frac{5\pi}{4} + i\sin\frac{5\pi}{4}\right)$$

7. Represent $\sqrt{3} + i$ graphically, and find the trigonometric form of the number.

Solution:

$$z = \sqrt{3} + i$$

$$r = \sqrt{(\sqrt{3})^2 + 1^2} = 2$$

$$\tan\theta = \frac{1}{\sqrt{3}}$$

$$\theta = 30° \text{ or } \frac{\pi}{6}$$

$$z = 2\left(\cos\frac{\pi}{6} + i\sin\frac{\pi}{6}\right)$$

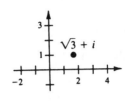

11. Represent $6i$ graphically, and find the trigonometric form of the number.

Solution:

$$z = 6i$$

$$r = \sqrt{0^2 + 6^2} = 6$$

$$\tan\theta = \frac{6}{0}, \text{ undefined}$$

$$\theta = \frac{\pi}{2}$$

$$z = 6\left(\cos\frac{\pi}{2} + i\sin\frac{\pi}{2}\right)$$

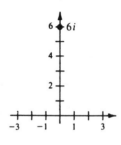

17. Represent $1 + 6i$ graphically, and find the trigonometric form of the number.

Solution:

$$z = 1 + 6i$$

$$r = \sqrt{37}$$

$$\tan\theta = 6$$

$$\theta \approx 1.41 \text{ rad}$$

$$z = \sqrt{37}[\cos(1.41) + i\sin(1.41)]$$

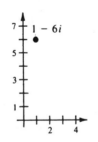

21. Represent $2(\cos 150° + i\sin 150°)$ graphically, and find the standard form of the number.

Solution:

$$z = 2(\cos 150° + i\sin 150°)$$

$$a = 2\cos 150° = -\sqrt{3}$$

$$b = 2\sin 150° = 1$$

$$z = -\sqrt{3} + i$$

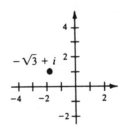

29. Represent $3[\cos(18°45') + i\sin(18°45')]$ graphically, and find the standard form of the number.

Solution:

$z = 3(\cos 18°45' + i\sin 18°45')$

$a = 3\cos 18°45' = 3\cos 18.75° = 2.8408$

$b = 3\sin 18°45' = 3\sin 18.75° = 0.9643$

$z = 2.8408 + 0.9643i$

2.84 + 0.96i

33. Perform the indicated operation and leave the result in trigonometric form.

$$\left[\tfrac{5}{3}(\cos 140° + i\sin 140°)\right]\left[\tfrac{2}{3}(\cos 60° + i\sin 60°)\right]$$

Solution:

$$\left[\tfrac{5}{3}(\cos 140° + i\sin 140°)\right]\left[\tfrac{2}{3}(\cos 60° + i\sin 60°)\right]$$
$$= \left(\tfrac{5}{3}\right)\left(\tfrac{2}{3}\right)[\cos(140° + 60°) + i\sin(140° + 60°)] = \tfrac{10}{9}[\cos 200° + i\sin 200°]$$

39. Perform the indicated operation and leave the result in trigonometric form.

$$\frac{\cos(5\pi/3) + i\sin(5\pi/3)}{\cos \pi + i\sin \pi}$$

Solution:

$$\frac{\cos(5\pi/3) + i\sin(5\pi/3)}{\cos \pi + i\sin \pi} = \cos\left(\frac{5\pi}{3} - \pi\right) + i\sin\left(\frac{5\pi}{3} - \pi\right) = \cos\frac{2\pi}{3} + i\sin\frac{2\pi}{3}$$

43. For $(2 + 2i)(1 - i)$, (a) give the trigonometric form of the complex numbers, (b) perform the indicated operation using the trigonometric form, and (c) perform the indicated operation using the standard form and check your result with the answer in part (b).

Solution:
(a) Trigonometric form:

$$\left[2\sqrt{2}\left(\cos\frac{\pi}{4} + i\sin\frac{\pi}{4}\right)\right]\left[\sqrt{2}\left(\cos\frac{7\pi}{4} + i\sin\frac{7\pi}{4}\right)\right]$$

(b) Operation in trigonometric form:

$$(2\sqrt{2})(\sqrt{2})\left[\cos\left(\frac{\pi}{4} + \frac{7\pi}{4}\right) + i\sin\left(\frac{\pi}{4} + \frac{7\pi}{4}\right)\right] = 4[\cos(2\pi) + i\sin(2\pi)]$$
$$= 4[\cos(0°) + i\sin(0°)]$$
$$= 4$$

(c) $(2 + 2i)(1 - i) = 2 - 2i + 2i - 2i^2 = 4$

47. For $5/(2 + 3i)$, (a) give the trigonometric form of the complex numbers, (b) perform the indicated operation using the trigonometric form, and (c) perform the indicated operation using the standard form and check your result with the answer in part (b).

Solution:

(a) Trigonometric form:

$$\frac{5[\cos 0° + i \sin 0°]}{\sqrt{13}[\cos 56.3° + i \sin 56.3°]}$$

(b) Operation in trigonometric form:

$$\frac{5}{\sqrt{13}}[\cos(-56.3°) + i \sin(-56.3°)] = \frac{5\sqrt{13}}{13}[\cos(-56.3°) + i \sin(-56.3°)] \approx \frac{5}{13}(2 - 3i)$$

(c) $\dfrac{5}{2 + 3i} \cdot \dfrac{2 - 3i}{2 - 3i} = \dfrac{5(2 - 3i)}{13} = \dfrac{5}{13}(2 - 3i)$

51. Use the trigonometric form $z = r(\cos \theta + i \sin \theta)$ and $\bar{z} = r[\cos(-\theta) + i \sin(-\theta)]$ to find (a) $z\bar{z}$ and (b) $z/\bar{z}$, $z \neq 0$.

Solution:

(a) $z\bar{z} = [r(\cos \theta + i \sin \theta)][r \cos(-\theta) + i \sin(-\theta)]$
$= r^2[\cos(\theta - \theta) + i \sin(\theta - \theta)]$
$= r^2$

(b) $\dfrac{z}{\bar{z}} = \dfrac{r}{r}[\cos(\theta - (-\theta)) + i \sin(\theta - (-\theta))]$
$= \cos 2\theta + i \sin 2\theta$

SECTION 5.4

DeMoivre's Theorem and nth Roots

- You should know DeMoivre's Theorem: If $z = r(\cos\theta + i\sin\theta)$, then for any positive integer n,

$$z^n = r^n(\cos n\theta + i\sin n\theta).$$

- You should know that for any positive integer n, $z = r(\cos\theta + i\sin\theta)$ has n distinct nth roots given by

$$\sqrt[n]{r}\left[\cos\left(\frac{\theta + 2\pi k}{n}\right) + i\sin\left(\frac{\theta + 2\pi k}{n}\right)\right]$$

where $k = 0, 1, 2, \ldots, n-1$.

Solutions to Selected Exercises

3. Use DeMoivre's Theorem to find $(-1+i)^{10}$. Express the result in standard form.

Solution:

$$
\begin{aligned}
(-1+i)^{10} &= \left[\sqrt{2}\left(\cos\frac{3\pi}{4} + i\sin\frac{3\pi}{4}\right)\right]^{10} \\
&= (\sqrt{2})^{10}\left[\cos 10\left(\frac{3\pi}{4}\right) + i\sin 10\left(\frac{3\pi}{4}\right)\right] \\
&= 32\left[\cos\frac{15\pi}{2} + i\sin\frac{15\pi}{2}\right] \\
&= 32[0 - i] \\
&= -32i
\end{aligned}
$$

7. Use DeMoivre's Theorem to find $[5(\cos 20° + i\sin 20°)]^3$. Express the result in standard form.

Solution:

$$
\begin{aligned}
[5(\cos 20° + i\sin 20°)]^3 &= 5^3[\cos 60° + i\sin 60°] \\
&= 125\left(\frac{1}{2} + \frac{\sqrt{3}}{2}i\right) \\
&= \frac{125}{2} + \frac{125\sqrt{3}}{2}i
\end{aligned}
$$

11. Use DeMoivre's Theorem to find $[5(\cos 3.2 + i \sin 3.2)]^4$. Express the result in standard form.

Solution:

$$[5(\cos 3.2 + i \sin 3.2)]^4 = 5^4[\cos 12.8 + i \sin 12.8]$$
$$\approx 625(0.97283 + 0.2315i)$$
$$\approx 608.02 + 144.69i$$

15. (a) Use DeMoivre's Theorem to find the fourth roots of

$$16\left(\cos \frac{4\pi}{3} + i \sin \frac{4\pi}{3}\right),$$

(b) represent each of the roots graphically, and (c) express each of the roots in standard form.

Solution:

(a) & (c) $n = 4$

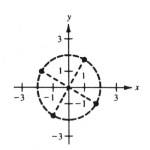

$$k = 0: \quad 2\left(\cos \frac{\pi}{3} + i \sin \frac{\pi}{3}\right) = 1 + \sqrt{3}i$$
$$k = 1: \quad 2\left(\cos \frac{5\pi}{6} + i \sin \frac{5\pi}{6}\right) = -\sqrt{3} + i$$
$$k = 2: \quad 2\left(\cos \frac{4\pi}{3} + i \sin \frac{4\pi}{3}\right) = -1 - \sqrt{3}i$$
$$k = 3: \quad 2\left(\cos \frac{11\pi}{6} + i \sin \frac{11\pi}{6}\right) = \sqrt{3} - i$$

19. (a) Use DeMoivre's Theorem to find the cube roots of

$$-\frac{125}{2}(1 + \sqrt{3}i),$$

(b) represent each of the roots graphically, and (c) express each of the roots in standard form.

Solution:

$$-\frac{125}{2}(1 + \sqrt{3}i) = 125\left(\cos \frac{4\pi}{3} + i \sin \frac{4\pi}{3}\right)$$

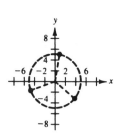

(a) & (c) $n = 3$

$$k = 0: \quad 5\left(\cos \frac{4\pi}{9} + i \sin \frac{4\pi}{9}\right) \approx 0.8682 + 4.9240i$$
$$k = 1: \quad 5\left(\cos \frac{10\pi}{9} + i \sin \frac{10\pi}{9}\right) \approx -4.6985 - 1.7101i$$
$$k = 2: \quad 5\left(\cos \frac{16\pi}{9} + i \sin \frac{16\pi}{9}\right) \approx 3.8302 - 3.2139i$$

25. Find all the solutions of $x^4 - i = 0$ and represent your solutions graphically.

Solution:

$$x^4 - i = 0$$
$$x^4 = i$$

Find the fourth roots of $i = \cos \dfrac{\pi}{2} + i \sin \dfrac{\pi}{2}$.

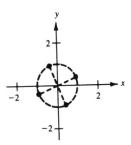

$n = 4$

$k = 0$: $\cos \dfrac{\pi}{8} + i \sin \dfrac{\pi}{8} \approx 0.9239 + 0.3827i$

$k = 1$: $\cos \dfrac{5\pi}{8} + i \sin \dfrac{5\pi}{8} \approx -0.3827 + 0.9239i$

$k = 2$: $\cos \dfrac{9\pi}{8} + i \sin \dfrac{9\pi}{8} \approx -0.9239 - 0.3827i$

$k = 3$: $\cos \dfrac{13\pi}{8} + i \sin \dfrac{13\pi}{8} \approx 0.3827 - 0.9239i$

29. Find all the solutions of $x^3 + 64i = 0$ and represent your solutions graphically.

Solution:

$$x^3 + 64i = 0$$
$$x^3 = -64i$$

Find the cube roots of $-64i = 64\left(\cos \dfrac{3\pi}{2} + i \sin \dfrac{3\pi}{2}\right)$.

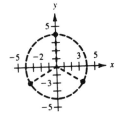

$n = 3$

$k = 0$: $4\left(\cos \dfrac{\pi}{2} + i \sin \dfrac{\pi}{2}\right) = 4i$

$k = 1$: $4\left(\cos \dfrac{7\pi}{6} + i \sin \dfrac{7\pi}{6}\right) = -2\sqrt{3} - 2i$

$k = 2$: $4\left(\cos \dfrac{11\pi}{6} + i \sin \dfrac{11\pi}{6}\right) = 2\sqrt{3} - 2i$

REVIEW EXERCISES FOR CHAPTER 5

Solutions to Selected Exercises

3. Perform the indicated operations and write the result in standard form.

$$\left(\frac{\sqrt{2}}{2} - \frac{\sqrt{2}}{2}i\right) - \left(\frac{\sqrt{2}}{2} + \frac{\sqrt{2}}{2}i\right)$$

Solution:

$$\left(\frac{\sqrt{2}}{2} - \frac{\sqrt{2}}{2}i\right) - \left(\frac{\sqrt{2}}{2} + \frac{\sqrt{2}}{2}i\right) = \left(\frac{\sqrt{2}}{2} - \frac{\sqrt{2}}{2}\right) + \left(-\frac{\sqrt{2}}{2} - \frac{\sqrt{2}}{2}\right)i$$

$$= 0 - \sqrt{2}i$$

$$= -\sqrt{2}i$$

7. Perform the indicated operations and write the result in standard form.

$$(10 - 8i)(2 - 3i)$$

Solution:

$$(10 - 8i)(2 - 3i) = 20 - 30i - 16i + 24i^2 = -4 - 46i$$

11. Perform the indicated operations and write the result in standard form.

$$\frac{4}{-3i}$$

Solution:

$$\frac{4}{-3i} = \frac{4}{-3i} \cdot \frac{3i}{3i} = \frac{12i}{9} = \frac{4i}{3} = 0 + \frac{4}{3}i$$

13. Use the discriminant to determine the number of real solutions of $6x^2 + x - 2 = 0$.

Solution:

$a = 6, \quad b = 1, \quad c = -2$

$b^2 - 4ac = (1)^2 - 4(6)(-2) = 49 > 0$

There are *two* real solutions.

17. Use the discriminant to determine the number of real solutions of $0.13x^2 - 0.45x + 0.65 = 0$.

Solution:

$a = 0.13, \quad b = -0.45, \quad c = 0.65$

$b^2 - 4ac = (-0.45)^2 - 4(0.13)(0.65) = -0.1355 < 0$

There are *no* real solutions.

21. Find a fourth degree polynomial with the zeros -1, -1, $\frac{1}{3}$, and $-\frac{1}{2}$.

Solution:

$$f(x) = 6(x+1)^2\left(x - \tfrac{1}{3}\right)\left(x + \tfrac{1}{2}\right) \qquad \text{Multiply by 6 to clear the fractions.}$$
$$= (x+1)^2 3\left(x - \tfrac{1}{3}\right)2\left(x + \tfrac{1}{2}\right)$$
$$= (x^2 + 2x + 1)(3x - 1)(2x + 1)$$
$$= (x^2 + 2x + 1)(6x^2 + x - 1)$$
$$= 6x^4 + 13x^3 + 7x^2 - x - 1$$

25. Find all the zeros of $f(x) = x^3 - 18x^2 + 106x - 200$. [*Hint:* One zero is $x = 7 + i$.]

Solution:

Since $7 + i$ is a zero, so is $7 - i$. Thus, $[x - (7 + i)]$ and $[x - (7 - i)]$ are factors of $f(x)$.

$$[x - (7+i)][x - (7-i)] = [(x - 7) - i][(x - 7) + i]$$
$$= (x - 7)^2 - i^2 = x^2 - 14x + 50$$

Now by long division, we have

$$
\begin{array}{r}
x \;-\; 4 \\
x^2 - 14x + 50 \overline{\smash{\big)}\, x^3 - 18x^2 + 106x - 200} \\
\underline{x^3 - 14x^2 + 50x} \\
-\;4x^2 + 56x - 200 \\
\underline{-\;4x^2 + 56x - 200} \\
0
\end{array}
$$

Therefore, $f(x) = (x^2 - 14x + 50)(x - 4)$ and the zeros of f are $x = 7 + i$, $7 - i$, 4.

29. Find all the zeros of $f(x) = x^4 + 5x^3 + 2x^2 - 50x - 84$. [*Hint:* One zero is $x = -3 + \sqrt{5}\,i$.]

Solution:

Since $x = -3 + \sqrt{5}\,i$ is a zero, so is $-3 - \sqrt{5}\,i$. Thus, $[x - (-3 + \sqrt{5}\,i)]$ and $[x - (-3 - \sqrt{5}\,i)]$ are factors of $f(x)$.

$$[x - (-3 + \sqrt{5}\,i)][x - (-3 - \sqrt{5}\,i)] = [(x + 3) - \sqrt{5}\,i][(x + 3) + \sqrt{5}\,i]$$
$$= (x + 3)^2 - 5i^2$$
$$= x^2 + 6x + 14$$

Now by long division, we have

$$
\begin{array}{r}
x^2 - x - 6 \\
x^2 + 6x + 14 \overline{\smash{)}\ x^4 + 5x^3 + 2x^2 - 50x - 84} \\
\underline{x^4 + 6x^3 + 14x^2} \\
-x^3 - 12x^2 - 50x \\
\underline{-x^3 - 6x^2 - 14x} \\
-6x^2 - 36x - 84 \\
\underline{-6x^2 - 36x - 84} \\
0
\end{array}
$$

Therefore, $f(x) = (x^2 + 6x + 14)(x^2 - x - 6) = (x^2 + 6x + 14)(x - 3)(x + 2)$ and the zeros of f are $x = -3 + \sqrt{5}\,i,\ -3 - \sqrt{5}\,i,\ 3,\ -2.$

33. Find the trigonometric form of $5 - 5i$.

Solution:

$$z = 5 - 5i$$
$$|z| = \sqrt{5^2 + (-5)^2} = 5\sqrt{2}$$
$$\tan\theta = -\frac{5}{5} = -1,\ \theta \text{ is in Quadrant IV}$$
$$\theta = 315°$$
$$z = 5\sqrt{2}(\cos 315° + i\sin 315°)$$

37. Write $100(\cos 240° + i\sin 240°)$ in standard form.

Solution:

$$z = 100(\cos 240° + i\sin 240°) = 100\left(-\frac{1}{2} - \frac{\sqrt{3}}{2}i\right) = -50 - 50\sqrt{3}\,i$$

43. Use DeMoivre's Theorem to find the indicated power of the following complex number. Express the result in standard form.

$$\left[5\left(\cos\frac{\pi}{12} + i\sin\frac{\pi}{12}\right)\right]^4$$

Solution:

$$\left[5\left(\cos\frac{\pi}{12} + i\sin\frac{\pi}{12}\right)\right]^4 = 625\left(\cos\frac{\pi}{3} + i\sin\frac{\pi}{3}\right) = \frac{625}{2} + \frac{625\sqrt{3}}{2}i$$

47. Use DeMoivre's Theorem to find the sixth roots of $-729i$.

Solution:

Find the sixth roots of $-729i = 729\left(\cos\dfrac{3\pi}{2} + i\sin\dfrac{3\pi}{2}\right)$.

$n = 6$

$k = 0:\ 3\left(\cos\dfrac{\pi}{4} + i\sin\dfrac{\pi}{4}\right)$

$k = 1:\ 3\left(\cos\dfrac{7\pi}{12} + i\sin\dfrac{7\pi}{12}\right)$

$k = 2:\ 3\left(\cos\dfrac{11\pi}{12} + i\sin\dfrac{11\pi}{12}\right)$

$k = 3:\ 3\left(\cos\dfrac{5\pi}{4} + i\sin\dfrac{5\pi}{4}\right)$

$k = 4:\ 3\left(\cos\dfrac{19\pi}{12} + i\sin\dfrac{19\pi}{12}\right)$

$k = 5:\ 3\left(\cos\dfrac{23\pi}{12} + i\sin\dfrac{23\pi}{12}\right)$

Practice Test for Chapter 5

1. Express i^{38} as i, $-i$, 1, or -1.

In Exercises 2–5, perform the indicated operation and write the result in standard form.

2. $\left(8 + \sqrt{-64}\right) + \left(6 + \sqrt{-25}\right)$

3. $-(4 + 4i) - (-3i)$

4. $(-8 + 2i)(-8 - 2i)$

5. $\dfrac{12 + 16i}{4 - 2i}$

6. Use the quadratic equation to solve $3x^2 + 2x + 2 = 0$.

In Exercises 7–10, find all the zeros of the given equation. List any complex solutions in $a + bi$ form.

7. $3x^2 + 1 = -47$

8. $x^4 - 1296 = 0$

9. $x^4 - 7x^2 - 60 = 0$

10. $x^3 + 2x^2 + 9x + 18 = 0$

11. Find a polynomial with integer coefficients that has the zeros 0, $-2i$, 3.

12. Find all the zeros of $f(x) = x^3 + 4x^2 - 7x + 30$ given that $r = 1 + 2i$ is a zero.

13. Give the trigonometric form of $z = 5 - 5i$.

14. Give the standard form of $z = 6(\cos 225° + i \sin 225°)$.

15. Multiply $[7(\cos 23° + i \sin 23°)][4(\cos 7° + i \sin 7°)]$.

16. Divide $\dfrac{9\left(\cos \dfrac{5\pi}{4} + i \sin \dfrac{5\pi}{4}\right)}{3(\cos \pi + i \sin \pi)}$.

17. Find $(2 + 2i)^8$.

18. Find the cube roots of $8\left(\cos \dfrac{\pi}{3} + i \sin \dfrac{\pi}{3}\right)$.

19. Find all the solutions to $x^3 + 125 = 0$.

20. Find all the solutions to $x^4 + i = 0$.

CHAPTER 6

Exponential and Logarithmic Functions

SECTION 6.1

Exponential Functions

- You should know that a function of the form $y = a^x$, where $a > 0$, $a \neq 1$, is called an exponential function with base a.

- You should be able to graph exponential functions.

- You should know some properties of exponential functions where $a > 0$ and $a \neq 1$.

  (a) If $a^x = a^y$, then $x = y$.

  (b) If $a^x = b^x$ and $x \neq 0$, then $a = b$.

- You should know formulas for compound interest.

  (a) For n compoundings per year: $A = P\left(1 + \dfrac{r}{n}\right)^{nt}$.

  (b) For continuous compoundings: $A = Pe^{rt}$.

Solutions to Selected Exercises

3. Use a calculator to evaluate $1000(1.06)^{-5}$. Round your answer to three decimal places.

Solution:

$$1000(1.06)^{-5} \approx 747.258$$

7. Use a calculator to evaluate $8^{2\pi}$. Round your answer to three decimal places.

Solution:

$$8^{2\pi} \approx 472,369.379$$

11. Use a calculator to evaluate e^2. Round your answer to three decimal places.

Solution:

$$e^2 \approx 7.389$$

15. Match $f(x) = 3^x$ with its graph.

Solution:
$f(x) = 3^x$
y-intercept: $(0, 1)$
3^x increases as x increases
Matches graph (g)

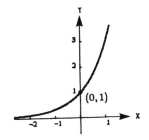

19. Match $f(x) = 3^x - 4$ with its graph.

Solution:
$f(x) = 3^x - 4$
y-intercept: $(0, -3)$
$3^x - 4$ increases as x increases
Matches graph (d)

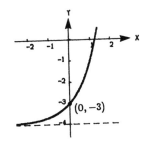

23. Sketch the graph of $g(x) = 5^x$.

Solution:
$g(x) = 5^x$

x	-2	-1	0	1	2
$g(x)$	$\frac{1}{25}$	$\frac{1}{5}$	1	5	25

27. Sketch the graph of $h(x) = 5^{x-2}$.

Solution:
$h(x) = 5^{x-2}$

x	-1	0	1	2	3
$h(x)$	$\frac{1}{125}$	$\frac{1}{25}$	$\frac{1}{5}$	1	5

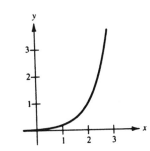

31. Sketch the graph of $y = 2^{-x^2}$.

Solution:

$$y = 2^{-x^2} = \left(\tfrac{1}{2}\right)^{x^2}$$

x	0	1	-1	2	-2
y	1	$\frac{1}{2}$	$\frac{1}{2}$	$\frac{1}{16}$	$\frac{1}{16}$

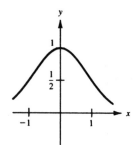

33. Sketch the graph of $y = 3^{|x|}$.

Solution:

$$y = 3^{|x|}$$

x	0	1	2	-1	-2
y	1	3	9	3	9

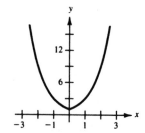

37. Sketch the graph of $f(x) = e^{2x}$.

Solution:

$$f(x) = e^{2x}$$

x	0	1	2	-1	-2
$f(x)$	1	7.39	54.60	0.135	0.02

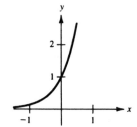

43. Complete the following table to determine the balance A for \$2500 invested at 12% for 20 years and compounded n times per year.

n	1	2	4	12	365	Continuous compounding
A						

Solution:

$$A = P\left(1 + \frac{r}{n}\right)^{nt}$$

$P = 2500, \ r = 0.12, \ t = 20$

When $n = 1, \ A = 2500\left(1 + \dfrac{0.12}{1}\right)^{(1)(20)} \approx \$24,115.73$

When $n = 2, \ A = 2500\left(1 + \dfrac{0.12}{2}\right)^{(2)(20)} \approx \$25,714.29$

When $n = 4, \ A = 2500\left(1 + \dfrac{0.12}{4}\right)^{(4)(20)} \approx \$26,602.23$

When $n = 12, \ A = 2500\left(1 + \dfrac{0.12}{12}\right)^{(12)(20)} \approx \$27,231.38$

When $n = 365, \ A = 2500\left(1 + \dfrac{0.12}{365}\right)^{(365)(20)} \approx \$27,547.07$

For continuous compounding, $A = Pe^{rt}, \quad A = 2500e^{(0.12)(20)} \approx \$27,557.94$

n	1	2	4	12	365	Continuous compounding
A	\$24,115.73	\$25,714.29	\$26,602.23	\$27,231.38	\$27,547.07	\$27,557.94

49. The demand equation for a certain product is given by $p = 500 - 0.5e^{0.004x}$. Find the price p for a demand of (a) $x = 1000$ units and (b) $x = 1500$ units.

Solution:
(a) $x = 1000$

$\quad p = 500 - 0.5e^4$

$\quad \approx \$472.70$

(b) $x = 1500$

$\quad p = 500 - 0.5e^6$

$\quad \approx \$298.29$

53. Given the exponential function $f(x) = a^x$, show that (a) $f(u + v) = f(u) \cdot f(v)$ and (b) $f(2x) = [f(x)]^2$.

Solution:
(a) $f(u + v) = a^{u+v}$

$\qquad\quad = a^u \cdot a^v$

$\qquad\quad = f(u) \cdot f(v)$

(b) $f(2x) = a^{2x}$

$\qquad\quad = (a^x)^2$

$\qquad\quad = [f(x)]^2$

SECTION 6.2

Logarithmic Functions

- You should know that a function of the form $y = \log_b M$, where $b > 0$, $b \neq 1$, and $M > 0$, is called a logarithm of M to base B.

- You should be able to convert from logarithmic form to exponential form and vice versa.

- You should know the following properties of logarithms.

  (a) $\log_a 1 = 0$

  (b) $\log_a a = 1$

  (c) $\log_a a^x = x$

- You should know the definition of the natural logarithmic function.

  $$\log_e x = \ln x, \quad x > 0$$

- You should know the properties of the natural logarithmic function.

  (a) $\ln 1 = 0$

  (b) $\ln e = 1$

  (c) $\ln e^x = x$

- You should know the change of base formula.

  $$\log_a x = \frac{\log_b x}{\log_b a}$$

- You should be able to graph logarithmic functions.

Solutions to Selected Exercises

5. Evaluate $\log_{16} 4$ without using a calculator.

   Solution:

   $$\log_{16} 4 = \log_{16} \sqrt{16} = \log_{16} 16^{1/2} = \tfrac{1}{2}$$

9. Evaluate $\log_{10} 0.01$ without using a calculator.

Solution:

$$\log_{10} 0.01 = \log_{10} \tfrac{1}{100} = \log_{10} 10^{-2} = -2$$

13. Evaluate $\ln e^{-2}$ without using a calculator.

Solution:

$$\ln e^{-2} = -2$$

17. Use the definition of a logarithm to write $5^3 = 125$ in logarithmic form.

Solution:

$$5^3 = 125$$
$$\log_5 125 = 3$$

23. Use the definition of a logarithm to write $e^3 = 20.0855\ldots$ in logarithmic form.

Solution:

$$e^3 = 20.0855\ldots$$
$$\log_e 20.0855\ldots = 3$$
$$\ln 20.0855\ldots \approx 3$$

27. Use a calculator to evaluate $\log_{10} 345$. Round your answer to three decimal places.

Solution:

$$\log_{10} 345 = 2.537819095\ldots$$
$$\approx 2.538$$

31. Use a calculator to evaluate $\ln(1 + \sqrt{3})$. Round your answer to three decimal places.

Solution:

$$\ln(1 + \sqrt{3}) = 1.005052539\ldots$$
$$\approx 1.005$$

35. Demonstrate that $f(x) = e^x$ and $g(x) = \ln x$ are inverses of each other by sketching their graphs on the same coordinate plane.

Solution:

x	-2	-1	0	1	2	3
$f(x)$	0.135	0.368	1	2.718	7.389	20.086
$g(x)$	—	—	—	0	0.693	1.097

The graph of g is obtained by reflecting the graph of f about the line $y = x$.

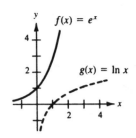

39. Use the graph of $y = \ln x$ to match $f(x) = -\ln(x + 2)$ to its graph.

Solution:
$f(x) = -\ln(x + 2)$
Vertical asymptote: $x = -2$
x-intercept: $(-1, \ 0)$
Matches graph (a)

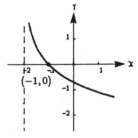

45. Find the domain, vertical asymptote, and x-intercept of $h(x) = \log_4(x - 3)$, and sketch its graph.

Solution:
Domain: $x - 3 > 0 \Rightarrow x > 3$
   The domain is $(3, \ \infty)$.
Vertical asymptote: $x - 3 = 0 \Rightarrow x = 3$
   The vertical asymptote is the line $x = 3$.
x-intercept: $\log_4(x - 3) = 0$
   $x - 3 = 1 \Rightarrow x = 4$
   The x-intercept is $(4, 0)$.

x	3.5	4	5	7
$h(x)$	-0.5	0	0.5	1

51. Use the change of base formula to write $\log_3 5$ as a multiple of a common logarithm.

Solution:

$$\log_3 5 = \frac{\log_{10} 5}{\log_{10} 3}$$

55. Use the change of base formula to write $\log_3 5$ as a multiple of a natural logarithm.

Solution:

$$\log_3 5 = \frac{\ln 5}{\ln 3}$$

59. Evaluate $\log_3 7$ using the change of base formula. Do the problem twice; once with common logarithms and once with natural logarithms. Round your answer to three decimal places.

Solution:

$$\log_3 7 = \frac{\log_{10} 7}{\log_{10} 3} = 1.771243749\ldots$$

$$\approx 1.771$$

$$\log_3 7 = \frac{\ln 7}{\ln 3} = 1.771243749\ldots$$

$$\approx 1.771$$

67. Students in a mathematics class were given an exam and then retested monthly with an equivalent exam. The average score for the class was given by the human memory model

$$f(t) = 80 - 17\log_{10}(t+1), \quad 0 \le t \le 12$$

where t is the time in months.

(a) What was the average score on the original exam $(t = 0)$?
(b) What was the average score after four months?
(c) What was the average score after ten months?

Solution:
(a) $f(0) = 80 - 17\log_{10} 1 = 80.0$
(b) $f(4) = 80 - 17\log_{10} 5 \approx 68.1$
(c) $f(10) = 80 - 17\log_{10} 11 \approx 62.3$

69. The population of a town will double in

$$t = \frac{10\ln 2}{\ln 67 - \ln 50}$$

years. Find t.

Solution:

$$t = \frac{10\ln 2}{\ln 67 - \ln 50}$$

$$t \approx \frac{6.931471806}{4.204692619 - 3.912023005}$$

$$t \approx 23.68 \text{ years}$$

73. (a) Use a calculator to complete the following table for the function

$$f(x) = \frac{\ln x}{x}.$$

x	1	5	10	10^2	10^4	10^6
$f(x)$						

(b) Use the table in part (a) to determine what $f(x)$ approaches as x increases without bound.

Solution:

(a)

x	1	5	10	10^2	10^4	10^6
$f(x)$	0	0.322	0.230	0.046	0.00092	0.0000138

(b) As $x \to \infty$, $f(x) \to 0$.

SECTION 6.3

Properties of Logarithms

■ You should know the following properties of logarithms.

(a) $\log_a(uv) = \log_a u + \log_a v$

(b) $\log_a(u/v) = \log_a u - \log_a v$

(c) $\log_a u^n = n \log_a u$

■ You should be able to rewrite logarithmic expressions.

Solutions to Selected Exercises

5. Use the properties of logarithms to write $\log_8 x^4$ as a sum, difference, or multiple of logarithms.

Solution:

$$\log_8 x^4 = 4 \log_8 x$$

9. Use the properties of logarithms to write $\log_2 xyz$ as a sum, difference, or multiple of logarithms.

Solution:

$$\log_2 xyz = \log_2[x(yz)] = \log_2 x + \log_2 yz = \log_2 x + \log_2 y + \log_2 z$$

15. Use the properties of logarithms to write the following expression as a sum, difference, or multiple of logarithms.

$$\log_b \frac{x^2}{y^2 z^3}$$

Solution:

$$\log_b \frac{x^2}{y^2 z^3} = \log_b x^2 - \log_b y^2 z^3 = \log_b x^2 - [\log_b y^2 + \log_b z^3] = 2 \log_b x - 2 \log_b y - 3 \log_b z$$

19. Use the properties of logarithms to write the following expression as a sum, difference, or multiple of logarithms.

$$\log_9 \frac{x^4 \sqrt{y}}{z^5}$$

Solution:

$$\log_9 \frac{x^4 \sqrt{y}}{z^5} = \log_9 x^4 \sqrt{y} - \log_9 z^5 = \log_9 x^4 + \log_9 \sqrt{y} - \log_9 z^5 = 4 \log_9 x + \tfrac{1}{2} \log_9 y - 5 \log_9 z$$

23. Write $\log_4 z - \log_4 y$ as the logarithm of a single quantity.

Solution:

$$\log_4 z - \log_4 y = \log_4 \frac{z}{y}$$

27. Write $\ln x - 3\ln(x+1)$ as the logarithm of a single quantity.

Solution:

$$\ln x - 3\ln(x+1) = \ln x - \ln(x+1)^3 = \ln \frac{x}{(x+1)^3}$$

33. Write $\ln x - 2[\ln(x+2) + \ln(x-2)]$ as the logarithm of a single quantity.

Solution:

$$\begin{aligned}
\ln x - 2[\ln(x+2) + \ln(x-2)] &= \ln x - 2\ln(x+2)(x-2) \\
&= \ln x - 2\ln(x^2-4) \\
&= \ln x - \ln(x^2-4)^2 \\
&= \ln \frac{x}{(x^2-4)^2}
\end{aligned}$$

37. Write $\frac{1}{3}[\ln y + 2\ln(y+4)] - \ln(y-1)$ as the logarithm of a single quantity.

Solution:

$$\begin{aligned}
\frac{1}{3}[\ln y + 2\ln(y+4)] - \ln(y-1) &= \frac{1}{3}[\ln y + \ln(y+4)^2] - \ln(y-1) \\
&= \frac{1}{3}\ln[y(y+4)^2] - \ln(y-1) \\
&= \ln \sqrt[3]{y(y+4)^2} - \ln(y-1) \\
&= \ln \frac{\sqrt[3]{y(y+4)^2}}{y-1}
\end{aligned}$$

41. Approximate $\log_b 6$ using the properties of logarithms, given $\log_b 2 \approx 0.3562$ and $\log_b 3 \approx 0.5646$.

Solution:

$$\log_b 6 = \log_b(2 \cdot 3) = \log_b 2 + \log_b 3 \approx 0.3562 + 0.5646 = 0.9208$$

47. Approximate $\log_b \sqrt{2}$ using the properties of logarithms, given $\log_b 2 \approx 0.3562$.

Solution:

$$\log_b \sqrt{2} = \log_b(2^{1/2}) = \tfrac{1}{2}\log_b 2 \approx \tfrac{1}{2}(0.3562) = 0.1781$$

51. Approximate $\log_b \frac{1}{4}$ using the properties of logarithms, given $\log_b 2 \approx 0.3562$.

Solution:

$$\log_b \tfrac{1}{4} = \log_b 1 - \log_b 4 = 0 - \log_b 2^2 = -2\log_b 2 \approx -2(0.3562) = -0.7124$$

55. Approximate the following using the properties of logarithms, given $\log_b 2 \approx 0.3562$ and $\log_b 3 \approx 0.5646$.

$$\log_b \left[\frac{(4.5)^3}{\sqrt{3}} \right]$$

Solution:

$$\log_b \left[\frac{(4.5)^3}{\sqrt{3}} \right] = 3\log_b 4.5 - \frac{1}{2}\log_b 3$$

$$= 3\log_b \frac{9}{2} - \frac{1}{2}\log_b 3$$

$$= 3[\log_b 9 - \log_b 2] - \frac{1}{2}\log_b 3$$

$$= 3[2\log_b 3 - \log_b 2] - \frac{1}{2}\log_b 3$$

$$= 3[2(0.5646) - 0.3562] - \frac{1}{2}(0.5646)$$

$$= 2.0367$$

59. Find the exact value of $\log_4 16^{1.2}$.

Solution:

$$\log_4 16^{1.2} = 1.2\log_4 16 = 1.2\log_4 4^2 = (1.2)(2)\log_4 4 = (2.4)(1) = 2.4$$

63. Use the properties of logarithms to simplify $\log_4 8$.

Solution:

$$\log_4 8 = \log_4 2^3 = 3\log_4 2 = 3\log_4 \sqrt{4} = 3\log_4 4^{1/2} = 3\left(\tfrac{1}{2}\right)\log_4 4 = \tfrac{3}{2}$$

67. Use the properties of logarithms to simplify $\log_5 \frac{1}{250}$.

Solution:

$$\log_5 \tfrac{1}{250} = \log_5 1 - \log_5 250 = 0 - \log_5 (125 \cdot 2)$$
$$= -\log_5 (5^3 \cdot 2) = -[\log_5 5^3 + \log_5 2]$$
$$= -[3\log_5 5 + \log_5 2] = -3 - \log_5 2$$

71. Prove that $\log_b \dfrac{u}{v} = \log_b u - \log_b v$.

Solution:

Let $x = \log_b u$ and $y = \log_b v$, then $b^x = u$ and $b^y = v$.

$$\frac{u}{v} = \frac{b^x}{b^y} = b^{x-y}$$

$$\log_b \left(\frac{u}{v} \right) = \log_b (b^{x-y}) = x - y = \log_b u - \log_b v$$

SECTION 6.4

Solving Exponential and Logarithmic Equations

■ You should be able to solve exponential and logarithmic equations.

■ To solve an exponential equation, take the logarithm of both sides.

■ To solve a logarithmic equation, rewrite it in exponential form.

Solutions to Selected Exercises

5. Solve $\left(\frac{3}{4}\right)^x = \frac{27}{64}$ for x.

Solution:

$$\left(\frac{3}{4}\right)^x = \frac{27}{64}$$
$$\left(\frac{3}{4}\right)^x = \left(\frac{3}{4}\right)^3$$
$$x = 3$$

9. Solve $\log_{10} x = -1$ for x.

Solution:

$$\log_{10} x = -1$$
$$x = 10^{-1} = \frac{1}{10}$$

13. Apply the inverse properties of $\ln x$ and e^x to simplify $e^{\ln(5x+2)}$.

Solution:

$$e^{\ln(5x+2)} = 5x + 2$$

17. Solve $10^x = 42$.

Solution:

$$10^x = 42$$
$$x = \log_{10} 42 \approx 1.6232$$

21. Solve $3(10^{x-1}) = 2$.

Solution:

$$3(10^{x-1}) = 2$$
$$10^{x-1} = \tfrac{2}{3}$$
$$x - 1 = \log_{10} \tfrac{2}{3}$$
$$x = 1 + \log_{10} \tfrac{2}{3} \approx 0.8239$$

25. Solve $500e^{-x} = 300$.

Solution:

$$500e^{-x} = 300$$
$$e^{-x} = \tfrac{3}{5}$$
$$-x = \ln \tfrac{3}{5}$$
$$x = -\ln \tfrac{3}{5} = \ln \tfrac{5}{3} \approx 0.5108$$

29. Solve $25e^{2x+1} = 962$.

Solution:

$$25e^{2x+1} = 962$$
$$e^{2x+1} = \tfrac{962}{25}$$
$$2x + 1 = \ln \tfrac{962}{25}$$
$$2x = -1 + \ln \tfrac{962}{25}$$
$$x = \tfrac{1}{2}\left[-1 + \ln \tfrac{962}{25}\right] = -\tfrac{1}{2} + \tfrac{1}{2} \ln \tfrac{962}{25} \approx 1.3251$$

35. Solve $\left(1 + \frac{0.10}{12}\right)^{12t} = 2$.

Solution:

$$\left(1 + \frac{0.10}{12}\right)^{12t} = 2$$
$$\ln\left(1 + \frac{0.10}{12}\right)^{12t} = \ln 2$$
$$12t \ln\left(1 + \frac{0.10}{12}\right) = \ln 2$$
$$t = \frac{\ln 2}{12 \ln\left(1 + \frac{0.10}{12}\right)} \approx 6.9603$$

39. Solve $\left(\frac{1}{1.0775}\right)^{N} = 0.2247$.

Solution:

$$\left(\frac{1}{1.0775}\right)^N = 0.2247$$

$$N = \frac{\ln(0.2247)}{\ln\left(\frac{1}{1.0775}\right)} = \frac{\ln(0.2247)}{\ln 1 - \ln(1.0775)} = \frac{\ln(0.2247)}{-\ln(1.0775)} \approx 20.0016$$

43. Solve $3(1 + e^{2x}) = 4$.

Solution:

$$3(1 + e^{2x}) = 4$$
$$1 + e^{2x} = \tfrac{4}{3}$$
$$e^{2x} = \tfrac{1}{3}$$
$$2x = \ln \tfrac{1}{3}$$
$$x = \tfrac{1}{2} \ln \tfrac{1}{3} \approx -0.5493$$

47. Solve $\dfrac{e^x + e^{-x}}{e^x - e^{-x}} = 2$.

Solution:

$$\frac{e^x + e^{-x}}{e^x - e^{-x}} = 2$$

$$\frac{e^x(e^x + e^{-x})}{e^x(e^x - e^{-x})} = 2$$

$$\frac{e^{2x} + 1}{e^{2x} - 1} = 2$$

$$e^{2x} + 1 = 2(e^{2x} - 1)$$

$$3 = e^{2x}$$

$$2x = \ln 3$$

$$x = \frac{1}{2} \ln 3 \approx 0.549$$

51. Solve $2 \ln x = 7$.

Solution:

$$2 \ln x = 7$$
$$\ln x = \tfrac{7}{2}$$
$$x = e^{7/2} \approx 33.1154$$

55. Solve $\log_{10}(z - 3) = 2$.

Solution:

$$\log_{10}(z - 3) = 2$$
$$z - 3 = 10^2$$
$$z = 10^2 + 3 = 103$$

59. Solve $\log_{10}(x + 4) - \log_{10} x = \log_{10}(x + 2)$.

Solution:

$$\log_{10}(x + 4) - \log_{10} x = \log_{10}(x + 2)$$

$$\log_{10}\left(\frac{x + 4}{4}\right) = \log_{10}(x + 2)$$

$$\frac{x + 4}{x} = x + 2$$

$$x + 4 = x^2 + 2x$$

$$0 = x^2 + x - 4$$

$$x = \frac{-1 \pm \sqrt{17}}{2} = -\frac{1}{2} \pm \frac{\sqrt{17}}{2} \qquad \text{Quadratic Formula}$$

63. Solve $\ln x^2 = (\ln x)^2$.

Solution:

$$\ln x^2 = (\ln x)^2$$

$$2 \ln x = (\ln x)^2$$

$$0 = (\ln x)^2 - 2 \ln x$$

$$0 = \ln x(\ln x - 2)$$

$$\ln x = 0 \qquad \text{or} \qquad \ln x - 2 = 0$$

$$x = e^0 \qquad \text{or} \qquad \ln x = 2$$

$$x = 1 \qquad \text{or} \qquad x = e^2$$

67. The demand equation for a certain product is given by $p = 500 - 0.5(e^{0.004x})$. Find the demand x for a price of (a) $p = \$350$ and (b) $p = \$300$.

Solution:

(a)
$$350 = 500 - 0.5(e^{0.004x})$$
$$-150 = -0.5(e^{0.004x})$$
$$300 = e^{0.004x}$$
$$0.004x = \ln 300$$
$$x = \frac{\ln 300}{0.004} \approx 1426 \text{ units}$$

(b)
$$300 = 500 - 0.5(e^{0.004x})$$
$$-200 = -0.5(e^{0.004x})$$
$$400 = e^{0.004x}$$
$$0.004x = \ln 400$$
$$x = \frac{\ln 400}{0.004} \approx 1498 \text{ units}$$

SECTION 6.5

Applications of Exponential and Logarithmic Functions

■ You should be able to solve compound interest problems.

(a) Compound interest formulas:

1. $A = P\left(1 + \dfrac{r}{n}\right)^{nt}$

2. $A = Pe^{rt}$

(b) Doubling time:

1. $t = \dfrac{\ln 2}{n\ln[1 + (r/n)]}$, n compoundings per year

2. $t = \dfrac{\ln 2}{r}$, continuous compounding

(c) Effective yield:

1. Effective yield $= \left(1 + \dfrac{r}{n}\right)^{n} - 1$, n compoundings per year

2. Effective yield $= e^{r} - 1$, continuous compounding

■ You should be able to solve growth and decay problems.

$$Q(t) = Ce^{kt}$$

(a) If $k > 0$, the population grows.

(b) If $k < 0$, the population decays.

(c) Ratio of Carbon 12 to Carbon 14 is $R(t) = \dfrac{1}{10^{12}}2^{-t/5700}$

■ You should be able to solve logistics model problems.

$$Q(t) = \dfrac{M}{1 + \left(\frac{M}{Q_0} - 1\right)e^{-kt}}$$

■ You should be able to solve intensity model problems.

$$S = k\log_{10}\dfrac{I}{I_0}$$

Solutions to Selected Exercises

5. $500 is deposited into an account with continuously compounded interest. If the balance is $1292.85 after ten years, find the annual percentage rate, the effective yield, and the time to double.

Solution:

$P = 500, \ A = 1292.85, \ t = 10$

$A = Pe^{rt}$

$$1292.85 = 500e^{10r}$$

$$\frac{1292.85}{500} = e^{10r}$$

$$10r = \ln\left(\frac{1292.85}{500}\right)$$

$$r = \frac{1}{10}\ln\left(\frac{1292.85}{100}\right) \approx 0.095 = 9.5\%$$

Effective yield $= e^{0.095} - 1 \approx 0.09966 \approx 9.97\%$

Time to double: $\quad 1000 = 500e^{0.095t}$

$$2 = e^{0.095t}$$

$$0.095t = \ln 2$$

$$t = \frac{\ln 2}{0.095} \approx 7.30 \text{ years}$$

9. $5000 is deposited into an account with continuously compounded interest. If the effective yield is 8.33%, find the annual percentage rate, the time to double, and the amount after 10 years.

Solution:

$P = 5000$

Effective yield $= 8.33\%$

$0.0833 = e^r - 1$

$$r = \ln 1.0833 \approx 0.0800 = 8\%$$

Time to double: $10,000 = 5000e^{0.08t}$

$$t = \frac{\ln 2}{0.08} \approx 8.66 \text{ years}$$

After 10 years: $A = 5000e^{0.08(10)} \approx \$11,127.70$

13. Determine the time necessary for $1000 to double if it is invested at 11% compounded (a) annually, (b) monthly, (c) daily, and (d) continuously.

Solution:

$P = 1000$, $r = 11\%$

(a) $n = 1$

$$t = \frac{\ln 2}{\ln(1 + 0.11)} \approx 6.642 \text{ years}$$

(b) $n = 12$

$$t = \frac{\ln 2}{12 \ln\left(1 + \frac{0.11}{12}\right)} \approx 6.330 \text{ years}$$

(c) $n = 365$

$$t = \frac{\ln 2}{365 \ln\left(1 + \frac{0.11}{365}\right)} \approx 6.302 \text{ years}$$

(d) Continuously

$$t = \frac{\ln 2}{0.11} \approx 6.301 \text{ years}$$

17. $50 is deposited monthly into a savings account at an annual rate of 7% compounded monthly. Find the balance, A, after 20 years given that

$$A = \frac{P(e^{rt} - 1)}{e^{r/12} - 1}.$$

Solution:

$p = 50$, $r = 7\%$, $t = 20$

$$A = \frac{50(e^{0.07(20)} - 1)}{e^{0.07/12} - 1} \approx \$26,111.12$$

21. The population P of a city is given by $P = 105,300e^{0.015t}$, where t is the time in years with $t = 0$ corresponding to 1985. According to this model, in what year will the city have a population of 150,000?

Solution:

$$150,000 = 105,300e^{0.015t}$$

$$0.015t = \ln\left(\frac{150,000}{105,300}\right)$$

$$t = \frac{1}{0.015} \ln\left(\frac{150,000}{105,300}\right) \approx 23.588 \text{ years}$$

$$1985 + 24 = 2009$$

The city will have a population of 150,000 in the year 2009.

25. The half-life of the isotope Ra^{226} is 1,620 years. If the initial quantity is 10 grams, how much will remain after 1000 years, and after 10,000 years?

Solution:

$$Q(t) = Ce^{kt}$$
$$Q = 10 \quad \text{when } t = 0 \Rightarrow 10 = Ce^0 \Rightarrow 10 = C$$
$$Q(t) = 10e^{kt}$$
$$Q = 5 \quad \text{when } t = 1620$$
$$5 = 10e^{1620k}$$
$$k = \tfrac{1}{1620} \ln\left(\tfrac{1}{2}\right)$$
$$Q(t) = 10e^{[\ln(1/2)/1620]t}$$

When $t = 1000$, $Q(t) = 10e^{[\ln(1/2)/1620](1000)} \approx 6.52$ grams.
When $t = 10,000$, $Q(t) = 10e^{[\ln(1/2)/1620](10000)} \approx 0.14$ gram.

29. The half-life of the isotope Pu^{230} is 24,360 years. If 2.1 grams remain after 1000 years, what is the initial quantity and how much will remain after 10,000 years?

Solution:

$$y = Ce^{[\ln(1/2)/24360]t}$$
$$2.1 = Ce^{[\ln(1/2)/24360](1000)}$$
$$C \approx 2.16$$

The initial quantity is 2.16 grams.
When $t = 10,000$, $y = 2.16e^{[\ln(1/2)/24360](10000)} \approx 1.62$ grams.

33. Find the constant k such that the exponential function $y = Ce^{kt}$ passes through the points $(0, 1)$ and $(4, 10)$.

Solution:

$$y = Ce^{kt}$$
$$1 = Ce^{k(0)}, \quad (0, 1)$$
$$1 = C$$
$$y = e^{kt}$$
$$10 = e^{4k}, \quad (4, 10)$$
$$4k = \ln 10$$
$$k = \frac{\ln 10}{4} \approx 0.5756$$

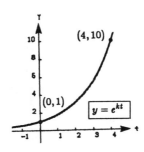

37. The sales S (in thousands of units) of a new product after it has been on the market t years are given by

$$S(t) = 100(1 - e^{kt}).$$

(a) Find S as a function of t if 15,000 units have been sold after one year.
(b) How many units will be sold after five years?

Solution:
$S(t) = 100(1 - e^{kt}), \quad S = 15$ when $t = 1$

(a) $15 = 100(1 - e^k)$

      $0.15 = 1 - e^k$

       $e^k = 0.85$

       $k = \ln 0.85$

     $S(t) = 100[1 - e^{(\ln 0.85)t}] = 100(1 - e^{-0.1625t})$

(b) $S(5) = 100[1 - e^{-(0.1625)(5)}] \approx 55.625$ thousands of units $= 55,625$ units

41. The intensity level β, in decibels, of a sound wave is defined by

$$\beta(I) = 10 \log_{10} \frac{I}{I_0}$$

where I_0 is an intensity of 10^{-16} watts per square centimeter, corresponding roughly to the faintest sound that can be heard. Determine $\beta(I)$ for the following conditions.

(a) $I = 10^{-14}$ watts per centimeter (whisper)
(b) $I = 10^{-9}$ watts per centimeter (busy street corner)
(c) $I = 10^{-6.5}$ watts per centimeter (air hammer)
(d) $I = 10^{-4}$ watts per centimeter (threshold of pain)

Solution:
$\beta(I) = 10 \log_{10} \dfrac{I}{I_0}$ where $I_0 = 10^{-16}$ watt/cm^2.

(a) $\beta(10^{-14}) = 10 \log_{10} \dfrac{10^{-14}}{10^{-16}} = 10 \log_{10} 10^2 = 20$

(b) $\beta(10^{-9}) = 10 \log_{10} \dfrac{10^{-9}}{10^{-16}} = 10 \log_{10} 10^7 = 70$

(c) $\beta(10^{-6.5}) = 10 \log_{10} \dfrac{10^{-6.5}}{10^{-16}} = 10 \log_{10} 10^{9.5} = 95$

(d) $\beta(10^{-4}) = 10 \log_{10} \dfrac{10^{-4}}{10^{-16}} = 10 \log_{10} 10^{12} = 120$

45. Use the acidity model $pH = -\log_{10}[H^+]$, where acidity (pH) is a measure of the hydrogen ion concentration $[H^+]$ (measured in moles of hydrogen per liter) of a solution. Find the pH if $[H^+] = 2.3 \times 10^{-5}$.

Solution:

$$pH = -\log_{10}[H^+] = -\log_{10}[2.3 \times 10^{-5}] \approx 4.64$$

49. Use **Newton's Law of Cooling** which states that the rate of change in the temperature of an object is proportional to the difference between its temperature and the temperature of its environment. If $T(t)$ is the temperature of the object at time t in minutes, T_0 is the initial temperature, and T_e is the constant temperature of the environment, then

$$T(t) = T_e + (T_0 - T_e)e^{-kt}.$$

An object in a room at 70° F cools from 350° F to 150° F in 45 minutes.

(a) Find the temperature of the object as a function of time.
(b) Find the temperature after it has cooled for one hour.
(c) Find the time necessary for the object to cool to 80° F.

Solution:

(a) $T_e = 70, \quad T_0 = 350, \quad T = 150$ when $t = 45$

$$150 = 70 + (350 - 70)e^{-45k}$$

$$80 = 280e^{-45k}$$

$$\frac{2}{7} = e^{-45k}$$

$$k = \frac{\ln(2/7)}{-45}$$

$$T(t) = 70 + 280e^{-[\ln(2/7)/-45]t} = 70 + 280e^{-0.02784t}$$

(b) $T(60) = 70 + 280e^{[\ln(2/7)/45](60)} \approx 122.7°$

(c) $\qquad 80 = 70 + 280e^{[\ln(2/7)/45]t}$

$$\frac{1}{28} = e^{[\ln(2/7)/45]t}$$

$$\frac{\ln(2/7)}{45}t = \ln\left(\frac{1}{28}\right)$$

$$t = \frac{45\ln(1/28)}{\ln(2/7)} \approx 119.7 \text{ minutes}$$

REVIEW EXERCISES FOR CHAPTER 6

Solutions to Selected Exercises

3. Sketch the graph of $g(x) = 6^{-x}$.

Solution:

$$g(x) = 6^{-x} = \left(\tfrac{1}{6}\right)^x$$

x	0	1	-1
$g(x)$	1	$\frac{1}{6}$	6

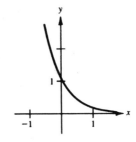

11. Sketch the graph of $f(x) = \ln x + 3$.

Solution:

$$f(x) = \ln x + 3$$

x	1	2	3	$\frac{1}{2}$	$\frac{1}{4}$
$f(x)$	3	3.69	4.10	2.31	1.61

15. Sketch the graph of $h(x) = \ln\left(e^{x-1}\right)$.

Solution:

$$\begin{aligned}
h(x) &= \ln\left(e^{x-1}\right) \\
&= (x-1)\ln e \\
&= x - 1
\end{aligned}$$

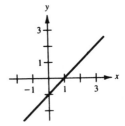

19. Use the properties of logarithms to write the following expression as a sum, difference, or multiple of logarithms.

$$\log_{10} \frac{5\sqrt{y}}{x^2}$$

Solution:

$$\begin{aligned}
\log_{10} \frac{5\sqrt{y}}{x^2} &= \log_{10} 5\sqrt{y} - \log_{10} x^2 \\
&= \log_{10} 5 + \log_{10} \sqrt{y} - \log_{10} x^2 \\
&= \log_{10} 5 + \frac{1}{2}\log_{10} y - 2\log_{10} x
\end{aligned}$$

23. Use the properties of logarithms to write the following expression as a sum, difference, or multiple of logarithms.

$$\ln[(x^2 + 1)(x - 1)]$$

Solution:

$$\ln[(x^2 + 1)(x - 1)] = \ln(x^2 + 1) + \ln(x - 1)$$

27. Write $\frac{1}{2} \ln |2x - 1| - 2 \ln |x + 1|$ as the logarithm of a single quantity.

Solution:

$$\frac{1}{2} \ln |2x - 1| - 2 \ln |x + 1| = \ln \sqrt{|2x - 1|} - \ln(x + 1)^2 = \ln \frac{\sqrt{|2x - 1|}}{(x + 1)^2}$$

31. Write $\ln 3 + \frac{1}{3} \ln(4 - x^2) - \ln x$ as the logarithm of a single quantity.

Solution:

$$\ln 3 + \frac{1}{3} \ln(4 - x^2) - \ln x = \ln 3 + \ln \sqrt[3]{4 - x^2} - \ln x = \ln(3 \sqrt[3]{4 - x^2}) - \ln x = \ln \frac{3 \sqrt[3]{4 - x^2}}{x}$$

35. Determine whether the equation $\ln(x + y) = \ln x + \ln y$ is true or false.

Solution:
False, since $\ln x + \ln y = \ln(xy)$.

39. Determine whether the following equation is true or false.

$$\frac{e^{2x} - 1}{e^x - 1} = e^x + 1$$

Solution:
True, since

$$\frac{e^{2x} - 1}{e^x - 1} = \frac{(e^x + 1)(e^x - 1)}{e^x - 1} = e^x + 1.$$

41. A solution of a certain drug contained 500 units per milliliter when prepared. It was analyzed after 40 days and found to contain 300 units per milliliter. Assuming that the rate of decomposition is proportional to the amount present, the equation giving the amount A after t days is

$$A = 500e^{-0.013t}.$$

Use this model to find A when $t = 60$.

Solution:

$$A = 500e^{-0.013(60)} \approx 229.2 \text{ units per milliliter}$$

45. A certain automobile gets 28 miles per gallon of gasoline for speeds up to 50 miles per hour. Over 50 miles per hour, the number of miles per gallon drops at the rate of 12% for each 10 miles per hour. If s is the speed and y is the number of miles per gallon, then

$$y = 28e^{0.6-0.012s}, \quad s \geq 50.$$

Use this function to complete the following table.

Speed	50	55	60	65	70
Miles per gallon					

Solution:

When $s = 50$, $y = 28e^{0.6-0.012(50)} = 28$ miles per gallon

When $s = 55$, $y = 28e^{0.6-0.012(55)} \approx 26.369$ miles per gallon

When $s = 60$, $y = 28e^{0.6-0.012(60)} \approx 24.834$ miles per gallon

When $s = 65$, $y = 28e^{0.6-0.012(65)} \approx 23.388$ miles per gallon

When $s = 70$, $y = 28e^{0.6-0.012(70)} \approx 22.026$ miles per gallon

Speed	50	55	60	65	70
Miles per gallon	28	26.4	24.8	23.4	22.0

49. Find the exponential function $y = Ce^{kt}$ that passes through the points $(0, 4)$ and $(5, \frac{1}{2})$.

Solution:

$$4 = Ce^{k(0)}, \quad (0, 4)$$
$$4 = C(1) \quad \text{so} \quad y = 4e^{kt}$$
$$\tfrac{1}{2} = 4e^{5k}, \quad (5, \tfrac{1}{2})$$
$$\tfrac{1}{8} = e^{5k}$$
$$5k = \ln \tfrac{1}{8}$$
$$k \approx -0.4159$$

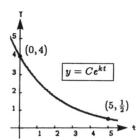

Thus, $y = 4e^{-0.4159t}$.

51. The demand equation for a certain product is given by

$$p = 500 - 0.5e^{0.004x}.$$

Find the demand x for a price of (a) $p = \$450$ and (b) $p = \$400$.

Solution:

(a)

$$p = 450$$
$$450 = 500 - 0.5e^{0.004x}$$
$$0.5e^{0.004x} = 50$$
$$e^{0.004x} = 100$$
$$0.004x = \ln 100$$
$$x \approx 1151 \text{ units}$$

(b)

$$p = 400$$
$$400 = 500 - 0.5e^{0.004x}$$
$$0.5e^{0.004x} = 100$$
$$e^{0.004x} = 200$$
$$0.004x = \ln 200$$
$$x \approx 1325 \text{ units}$$

57. In calculus it can be shown that

$$e^x \approx 1 + x + \frac{x^2}{2} + \frac{x^3}{6} + \frac{x^4}{24}.$$

Use this equation to approximate the following and compare the results to those obtained with a calculator.

(a) e (b) $e^{1/2}$ (c) $e^{-1/2}$

Solution:

(a) $e = e^1 \approx 1 + 1 + \frac{1}{2} + \frac{1}{6} + \frac{1}{24} \approx 2.7083$
By calculator: $e \approx 2.7183$

(b) $e^{1/2} \approx 1 + \frac{1}{2} + \frac{1/4}{2} + \frac{1/8}{6} + \frac{1/16}{24} \approx 1.6484$
By calculator: $e^{1/2} \approx 1.6487$

(c) $e^{-1/2} \approx 1 - \frac{1}{2} + \frac{1/4}{2} - \frac{1/8}{6} + \frac{1/16}{24} \approx 0.6068$
By calculator: $e^{-1/2} \approx 0.6065$

Practice Test for Chapter 6

1. Solve for x: $x^{3/5} = 8$.

2. Solve for x: $3^{x-1} = \frac{1}{81}$.

3. Graph $f(x) = 2^{-x}$.

4. Graph $g(x) = e^x + 1$.

5. If $5000 is invested at 9% interest, find the amount after three years if the interest is compounded
   (a) monthly   (b) quarterly   (c) continuously.

6. Write the equation in logarithmic form: $7^{-2} = \frac{1}{49}$.

7. Solve for x : $x - 4 = \log_2 \frac{1}{64}$.

8. Given $\log_b 2 = 0.3562$ and $\log_b 5 = 0.8271$, evaluate $\log_b \sqrt[4]{8/25}$.

9. Write $5 \ln x - \frac{1}{2} \ln y + 6 \ln z$ as a single logarithm.

10. Using your calculator and the change of base formula, evaluate $\log_9 28$.

11. Use your calculator to solve for N : $\log_{10} N = 0.6646$.

12. Graph $y = \log_4 x$.

13. Determine the domain of $f(x) = \log_3 (x^2 - 9)$.

14. Graph $y = \ln(x - 2)$.

15. True or False: $\dfrac{\ln x}{\ln y} = \ln(x - y)$.

16. Solve for x : $5^x = 41$.

17. Solve for x : $x - x^2 = \log_5 \frac{1}{25}$.

18. Solve for x : $\log_2 x + \log_2 (x - 3) = 2$.

19. Solve for x : $\dfrac{e^x + e^{-x}}{3} = 4$.

20. $6000 is deposited into a fund at an annual percentage rate of 13%. Find the time required for the investment to double if the interest is compounded continuously.

CHAPTER 7

Lines in the Plane and Conics

SECTION 7.1

Slope

Solutions to Selected Exercises

9. Plot the points $(-3, -2)$ and $(1, 6)$ and find the slope of the line passing through the points.

Solution:

$$m = \frac{6 - (-2)}{1 - (-3)} = \frac{8}{4} = 2$$

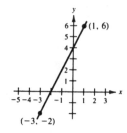

11. Plot the points $(-6, -1)$ and $(-6, 4)$ and find the slope of the line passing through the points.

Solution:

$$m = \frac{4 - (-1)}{-6 - (-6)} = \frac{5}{0}$$

The slope is undefined.

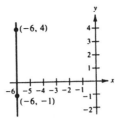

15. Use the point $(2, 1)$ on the line and the slope $m = 0$ of the line to find three additional points that the line passes through. (The solution is not unique.)

Solution:

Since $m = 0$, the line is horizontal, and since the line passes through $(2, 1)$, all other points on the line will be of the form $(x, 1)$. Three additional points are: $(0, 1)$, $(1, 1)$, $(3, 1)$.

21. Find the slope and y-intercept, if possible, of the line specified by $5x - y + 3 = 0$.

Solution:

$$5x - y + 3 = 0$$
$$-y = -5x - 3$$
$$y = 5x + 3$$

Slope: $m = 5$
y-intercept: $(0, 3)$

23. Find the slope and y-intercept, if possible, of the line specified by $5x - 2 = 0$.

Solution:

$$5x - 2 = 0$$
$$5x = 2$$
$$x = \tfrac{2}{5} \qquad \text{Vertical line}$$

Slope: Undefined
y-intercept: None

29. Find an equation for the line passing through the points $\left(2, \tfrac{1}{2}\right)$, $\left(\tfrac{1}{2}, \tfrac{5}{4}\right)$ and sketch a graph of the line.

Solution:

$$m = \frac{\frac{5}{4} - \frac{1}{2}}{\frac{1}{2} - 2} = \frac{\frac{3}{4}}{-\frac{3}{2}} = -\frac{1}{2}$$

$$y - \frac{1}{2} = -\frac{1}{2}(x - 2)$$

$$2y - 1 = -(x - 2)$$

$$2y - 1 = -x + 2$$

$$x + 2y - 3 = 0$$

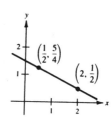

33. Find an equation for the line passing through the points $(1, 0.6)$, and $(-2, -0.6)$ and sketch a graph of the line.

Solution:

$$m = \frac{-0.6 - 0.6}{-2 - 1} = 0.4$$

$$y - 0.6 = 0.4(x - 1)$$

$$y - 0.6 = 0.4x - 0.4$$

$$y = 0.4x + 0.2 \quad \text{or} \quad 2x - 5y + 1 = 0$$

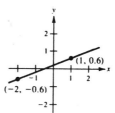

37. Find an equation of the line that passes through the point $(-3, 6)$ and has a slope of $m = -2$. Sketch a graph of the line.

Solution:

$$y - 6 = -2(x - (-3))$$

$$y - 6 = -2x - 6$$

$$y = -2x$$

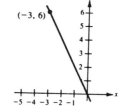

41. Find an equation of the line that passes through the point $(6, -1)$ and has an undefined slope. Sketch a graph of the line.

Solution:
Since the slope is undefined, the line is vertical and since the line passes through $(6, -1)$, its equation is $x = 6$.

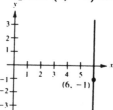

45. Prove that the line with intercepts $(a, 0)$ and $(0, b)$ has the following equation.

$$\frac{x}{a} + \frac{y}{b} = 1, \quad a \neq 0, \ b \neq 0$$

Solution:
Using the points $(a, 0)$ and $(0, b)$ we have

$$m = \frac{b-0}{0-a} = -\frac{b}{a}$$

$$y - 0 = -\frac{b}{a}(x-a)$$

$$y = -\frac{b}{a}x + b$$

$$ay = -bx + ab$$

$$bx + ay = ab$$

$$\frac{bx + ay}{ab} = \frac{ab}{ab}$$

$$\frac{x}{a} + \frac{y}{b} = 1.$$

49. Use the result of Exercise 45 to write an equation of the line with x-intercept $\left(-\frac{1}{6}, 0\right)$ and y-intercept $\left(0, -\frac{2}{3}\right)$.

Solution:

x-intercept: $\left(-\frac{1}{6}, 0\right)$

y-intercept: $\left(0, -\frac{2}{3}\right)$

$$\frac{x}{-1/6} + \frac{y}{-2/3} = 1$$

$$-6x - \frac{3}{2}y = 1$$

$$-12x - 3y = 2$$

$$12x + 3y = -2$$

53. Find the equation of the line giving the relationship between the temperature in degrees Celsius, C, and degrees Fahrenheit, F. Use the fact that water freezes at $0°$ Celsius ($32°$ Fahrenheit) and boils at $100°$ Celsius ($212°$ Fahrenheit).

Solution:

Using the points $(0, 32)$ and $(100, 212)$, we have

$$m = \frac{212 - 32}{100 - 0} = \frac{180}{100} = \frac{9}{5}$$

$$F - 32 = \frac{9}{5}(C - 0)$$

$$F = \frac{9}{5}C + 32$$

57. A store is offering a 15% discount on all items in its inventory. Write a linear equation giving the sale price S, for an item with a list price, L.

Solution:

$$S = L - 0.15L$$

$$S = 0.85L$$

SECTION 7.2

Additional Properties of Lines

- Two distinct nonvertical lines are *parallel* if and only if their slopes are equal.

- Two nonvertical lines are *perpendicular* if and only if

$$m_1 = -\frac{1}{m_2}.$$

- The distance between the point $(x_1,\ y_1)$ and the line $Ax + By + C = 0$ is

$$d = \frac{|Ax_1 + By_1 + C|}{\sqrt{A^2 + B^2}}.$$

- If a line has inclination θ and slope m, then

$$m = \tan\theta.$$

- The angle θ between two nonperpendicular lines is given by

$$\tan\theta = \left|\frac{m_2 - m_1}{1 + m_1 m_2}\right|.$$

Solutions to Selected Exercises

1. Determine if the lines L_1 and L_2 passing through the given pairs of points are parallel, perpendicular, or neither.

$$L_1 : (0,\ -1),\ (5,\ 9)$$
$$L_2 : (0,\ 3),\ (4,\ 1)$$

Solution:

The slope of L_1 is $m_1 = \dfrac{9 - (-1)}{5 - 0} = \dfrac{10}{5} = 2.$

The slope of L_2 is $m_2 = \dfrac{1 - 3}{4 - 0} = -\dfrac{2}{4} = -\dfrac{1}{2}.$

Since $m_1 \cdot m_2 = 2\left(-\frac{1}{2}\right) = -1$, the lines are perpendicular.

7. Write the equation of the line through the point $(-6,\ 4)$ (a) parallel to the line $3x + 4y = 7$ and (b) perpendicular to the line $3x + 4y = 7$.

Solution:

$$3x + 4y = 7$$
$$4y = -3x + 7$$
$$y = -\frac{3}{4}x + \frac{7}{4}$$

The slope of the given line is $m_1 = -3/4$.

(a) The slope of the parallel line is $m_2 = m_1 = -3/4$.

$$y - 4 = -\frac{3}{4}(x - (-6))$$
$$y - 4 = -\frac{3}{4}x - \frac{9}{2}$$
$$y = -\frac{3}{4}x - \frac{1}{2}$$
$$4y = -3x - 2$$
$$3x + 4y = -2$$

(b) The slope of the perpendicular line is $m_2 = -1/m_1 = 4/3$.

$$y - 4 = \frac{4}{3}(x - (-6))$$
$$y - 4 = \frac{4}{3}x + 8$$
$$y = \frac{4}{3}x + 12$$
$$3y = 4x + 36$$
$$4x - 3y = -36$$

13. Determine whether the three points, $(1,\ -3)$, $(3,\ 2)$, $(-2,\ 4)$, are vertices of a right triangle.

Solution:
The slope of the line connecting $(-2,\ 4)$ and $(3,\ 2)$ is

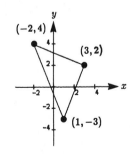

$$m_1 = \frac{2 - 4}{3 + 2} = -\frac{2}{5}$$

and the slope of the line connecting $(3,\ 2)$ and $(1,\ -3)$ is

$$m_2 = \frac{-3 - 2}{1 - 3} = \frac{5}{2}.$$

Since $m_1 = -1/m_2$, the two sides of the triangle are perpendicular. Thus, the triangle is a right triangle.

17. Determine whether the three points, $(0, -4)$, $(2, 0)$, $(3, 2)$, are collinear (lie on the same line).

Solution:

Let m_1 be the slope of the line through the points $(0, -4)$ and $(2, 0)$, and let m_2 be the slope of the line through the points $(2, 0)$ and $(3, 2)$.

$$m_1 = \frac{0 - (-4)}{2 - 0} = 2$$

$$m_2 = \frac{2 - 0}{3 - 2} = 2$$

Since $m_1 = m_2$, the points are collinear.

21. Find the distance between the point $(0, 0)$ and the line $4x + 3y = 10$.

Solution:

$4x + 3y - 10 = 0$

$A = 4$, $B = 3$, $C = -10$, $x_1 = y_1 = 0$

$$\frac{|Ax_1 + By_1 + C|}{\sqrt{A^2 + B^2}} = \frac{|4(0) + 3(0) + (-10)|}{\sqrt{(4)^2 + (3)^2}} = \frac{10}{5} = 2$$

25. Find the distance between the point $\left(\frac{1}{2}, \frac{2}{3}\right)$ and the line $y - 2 = 0$.

Solution:

$A = 0$, $B = 1$, $C = -2$, $x_1 = \frac{1}{2}$, $y_1 = \frac{2}{3}$

$$\frac{|Ax_1 + By_1 + C|}{\sqrt{A^2 + B^2}} = \frac{|0\left(\frac{1}{2}\right) + 1\left(\frac{2}{3}\right) + (-2)|}{\sqrt{(0)^2 + (1)^2}} = \frac{4}{3}$$

29. Find the distance between the lines $x + y = 1$ and $x + y = 5$.

Solution:

$(1, 0)$ is a point on the first line. The distance between $(1, 0)$ and $x + y - 5 = 0$ is

$$\frac{|1(1) + 1(0) + (-5)|}{\sqrt{(1)^2 + (1)^2}} = \frac{4}{\sqrt{2}} = 2\sqrt{2}$$

31. Find the inclination of the line $x - \sqrt{3}\, y = 0$.

Solution:

$$m = \frac{1}{\sqrt{3}}$$

$$\tan \theta = \frac{1}{\sqrt{3}} \Rightarrow \theta = 30°$$

41. Find the angle between the lines $4x + 3y + 2 = 0$ and $3x + 4y - 7 = 0$.

Solution:

The slope of $4x + 3y + 2 = 0$ is $m_1 = -\frac{4}{3}$.

The slope of $3x + 4y - 7 = 0$ is $m_2 = -\frac{3}{4}$.

$$\tan \theta = \frac{m_2 - m_1}{1 + m_1 m_2} = \frac{-\frac{3}{4} - \left(-\frac{4}{3}\right)}{1 + \left(-\frac{4}{3}\right)\left(-\frac{3}{4}\right)} = \frac{\frac{7}{12}}{2} = \frac{7}{24}$$

$$\theta = \arctan \frac{7}{24} \approx 16.3°$$

45. Find an equation of the line with the inclination $\theta = 60°$ and passing through the point $(0, 6)$.

Solution:

$$m = \tan 60° = \sqrt{3}$$
$$y - 6 = \sqrt{3}(x - 0)$$
$$y - 6 = \sqrt{3}\, x$$
$$y = \sqrt{3}\, x + 6 \quad \text{or} \quad y - \sqrt{3}\, x = 6$$

SECTION 7.3

Introduction to Conics: Parabolas

- A *parabola* is the set of all points (x, y) that are equidistant from a fixed line (*directrix*) and a fixed point (*focus*) not on the line.

- The standard equation of a parabola with vertex (h, k) and:

  (a) Vertical axis $x = h$ and directrix $y = k - p$ is:

  $$(x - h)^2 = 4p(y - k)$$

  (b) Horizontal axis $y = k$ and directrix $x = h - p$ is:

  $$(y - k)^2 = 4p(x - h)$$

Solutions to Selected Exercises

5. Match $(y - 1)^2 = 4(x - 2)$ with the correct graph.

Solution:
The vertex is at $(2, 1)$ and the axis is horizontal. Therefore, it matches graph (d).

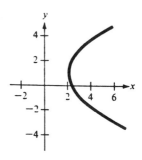

9. Find the vertex, focus, and directrix of the parabola $y^2 = -6x$.

Solution:

$$y^2 = -6x$$
$$(y - 0)^2 = 4\left(-\tfrac{3}{2}\right)(x - 0)$$

$h = 0, \ k = 0, \ p = -\tfrac{3}{2}$

Vertex: $(0, 0)$

Focus: $\left(-\tfrac{3}{2}, 0\right)$

Directrix: $x = 0 - \left(-\tfrac{3}{2}\right) = \tfrac{3}{2}$

13. Find the vertex, focus, and directrix of the parabola $(x-1)^2 + 8(y+2) = 0$.

Solution:

$$(x-1)^2 + 8(y+2) = 0$$
$$(x-1)^2 = -8(y+2)$$
$$(x-1)^2 = 4(-2)(y+2)$$

$h = 1,\ k = -2,\ p = -2$
Vertex: $(1,\ -2)$
Focus: $(1,\ -2+(-2))$ OR $(1,\ -4)$
Directrix: $y = -2-(-2)$ OR $y = 0$

17. Find the vertex, focus, and directrix of the parabola $y = \frac{1}{4}(x^2 - 2x + 5)$.

Solution:

$$y = \frac{1}{4}(x^2 - 2x + 5)$$
$$4y = x^2 - 2x + 1 - 1 + 5$$
$$4y = (x-1)^2 + 4$$
$$4y - 4 = (x-1)^2$$
$$(x-1)^2 = 4(y-1)$$
$$(x-1)^2 = 4(1)(y-1)$$

$h = 1,\ k = 1,\ p = 1$
Vertex: $(1,\ 1)$
Focus: $(1,\ 1+1)$ OR $(1,\ 2)$
Directrix: $y = 1-1$ OR $y = 0$

21. Find the vertex, focus, and directrix of the parabola $y^2 + 6y + 8x + 25 = 0$.

Solution:

$$y^2 + 6y + 8x + 25 = 0$$
$$y^2 + 6y = -8x - 25$$
$$y^2 + 6y + 9 = -8x - 25 + 9$$
$$(y+3)^2 = -8x - 16$$
$$(y+3)^2 = -8(x+2)$$
$$(y+3)^2 = 4(-2)(x+2)$$

$h = -2,\ k = -3,\ p = -2$
Vertex: $(-2,\ -3)$
Focus: $(-2+(-2),\ -3)$ OR $(-4,\ -3)$
Directrix: $x = -2-(-2)$ OR $x = 0$

25. Find the vertex, focus, and directrix of the parabola $x^2 + 4x + 4y - 4 = 0$.

Solution:

$$x^2 + 4x + 4y - 4 = 0$$
$$x^2 + 4x = -4y + 4$$
$$x^2 + 4x + 4 = -4y + 4 + 4$$
$$(x + 2)^2 = -4y + 8$$
$$(x + 2)^2 = -4(y - 2)$$
$$(x + 2)^2 = 4(-1)(y - 2)$$

$h = -2, \; k = 2, \; p = -1$

Vertex: $(-2, \; 2)$

Focus: $(-2, \; 2 + (-1))$ OR $(-2, \; 1)$

Directrix: $y = 2 - (-1)$ OR $y = 3$

29. Find an equation of the specified parabola.

Vertex: $(0, \; 0)$
Focus: $(-2, \; 0)$

Solution:

The axis is horizontal with $p = -2$.

$$(y - 0)^2 = 4(-2)(x - 0)$$
$$y^2 = -8x$$

33. Find an equation of the specified parabola.

Vertex: $(3, \; 2)$
Focus: $(1, \; 2)$

Solution:

The axis is horizontal with $p = -2$.

$$(y - 2)^2 = 4(-2)(x - 3)$$
$$y^2 - 4y + 4 = -8x + 24$$
$$y^2 - 4y + 8x - 20 = 0$$

37. Find an equation of the specified parabola.

Focus: $(0, \; 0)$
Directrix: $y = 4$

Solution:

The vertex is $(0, 2)$. The axis is vertical with $p = -2$.

$$(x - 0)^2 = 4(-2)(y - 2)$$
$$x^2 = -8y + 16$$
$$x^2 + 8y - 16 = 0$$

41. Find an equation of the specified parabola.

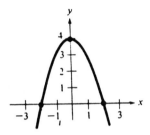

Solution:

The x-intercepts occur at $(\pm 2, 0)$ and the parabola opens downward.

$$y = -(x + 2)(x - 2)$$
$$y = -(x^2 - 4)$$
$$y = 4 - x^2$$

45. The receiver in a parabolic television dish antenna is 3 feet from the vertex and is located at the focus, as shown in the figure. Find an equation of a cross section of the reflector. (Assume the dish is directed upward and the vertex is at the origin.)

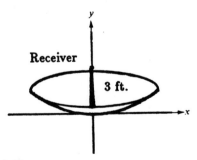

Receiver

3 ft.

Solution:

The vertex is at $(0, 0)$ and the focus is at $(3, 0)$. Therefore, $p = 3$.

$$(x - 0)^2 = 4(3)(y - 0)$$
$$x^2 = 12y$$

49. Find the equation of the tangent line to the parabola $y = -2x^2$ at the point $(-1, -2)$.

Solution:

$$y = -2x^2$$

$$x^2 = -\frac{1}{2}y$$

$$x^2 = 4\left(-\frac{1}{8}\right)y \Rightarrow p = -\frac{1}{8}$$

The focus is at $\left(-\frac{1}{8}, 0\right)$.

$$d_1 = \frac{1}{8} + b$$

$$d_2 = \sqrt{(-1-0)^2 + \left(-2 - \left(-\frac{1}{8}\right)\right)^2} = \sqrt{1 + \frac{225}{64}} = \frac{17}{4}$$

$$d_1 = d_2$$

$$\frac{1}{8} + b = \frac{17}{8}$$

$$b = 2$$

$$m = \frac{-2 - 2}{-1 - 0} = 4$$

$$y = 4x + 2$$

53. A ball is thrown horizontally from the top of a 75-foot tower with a velocity of 32 feet per second.

(a) Find the equation of the parabolic path.

(b) How far does the ball travel horizontally before striking the ground?

Solution:

(a) $y = -16t^2 + v_0 t + s_0$

$y = -16t^2 + 75$

Letting $x = 32t$, we have $t = x/32$ and

$$y = -16\left(\frac{x}{32}\right)^2 + 75$$

$$y = -\frac{x^2}{64} + 75 \quad \text{or} \quad x^2 + 64y = 4800$$

(b) When $y = 0$, we have

$$x^2 = 4800$$

$$x = \sqrt{4800} = 40\sqrt{3} \approx 69.28 \text{ feet away}$$

SECTION 7.4

Ellipses

- An *ellipse* is the set of all points (x, y) the sum of whose distances from two distinct points (*foci*) is constant.

- The standard equation of an ellipse with center (h, k) and major and minor axes of lengths $2a$ and $2b$ is:

  (a) $\dfrac{(x + h)^2}{a^2} + \dfrac{(y - k)^2}{b^2} = 1$ if the major axis is horizontal.

  (b) $\dfrac{(x - h)^2}{b^2} + \dfrac{(y - k)^2}{a^2} = 1$ if the major axis is vertical.

- $c^2 = a^2 - b^2$ where c is the distance from the center to a focus.

- The eccentricity of an ellipse is $e = \dfrac{c}{a}$.

Solutions to Selected Exercises

3. Match the following equation with the correct graph.

$$\frac{x^2}{9} + \frac{y^2}{4} = 1$$

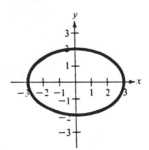

Solution:
$a = 3, \ b = 2$
Center: $(0, 0)$
Major axis is horizontal
Therefore, it matches graph (c).

7. Find the center, foci, vertices, and eccentricity of the following ellipse and sketch its graph.

$$\frac{x^2}{25} + \frac{y^2}{16} = 1$$

Solution:

$a^2 = 25, \; b^2 = 16, \; c^2 = 9$

Center: $(0, 0)$

Foci: $(\pm 3, 0)$

Vertices: $(\pm 5, 0), \quad e = \frac{3}{5}$

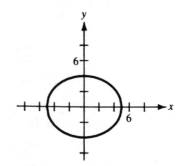

11. Find the center, foci, vertices, and eccentricity of the following ellipse and sketch its graph.

$$\frac{x^2}{9} + \frac{y^2}{5} = 1$$

Solution:

$a^2 = 9, \; b^2 = 5, \; c^2 = 4$

Center: $(0, 0)$

Foci: $(\pm 2, 0)$

Vertices: $(\pm 3, 0), \quad e = \frac{2}{3}$

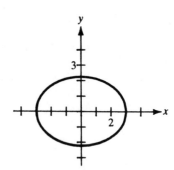

15. Find the center, foci, vertices, and eccentricity of the ellipse $3x^2 + 2y^2 = 6$ and sketch its graph.

Solution:

$$3x^2 + 2y^2 = 6$$

$$\frac{x^2}{2} + \frac{y^2}{3} = 1$$

$a^2 = 3, \; b^2 = 2, \; c^2 = 1$

Center: $(0, 0)$

Foci: $(0, \pm 1)$

Vertices: $\left(0, \pm\sqrt{3}\right), \quad e = \frac{1}{\sqrt{3}} = \frac{\sqrt{3}}{3}$

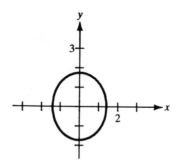

19. Find the center, foci, vertices, and eccentricity of the following ellipse and sketch its graph.

$$\frac{(x-1)^2}{9} + \frac{(y-5)^2}{25} = 1$$

Solution:
$a^2 = 25, \ b^2 = 9, \ c^2 = 16$
Center: $(1, 5)$
Foci: $(1, 9), \ (1, 1)$
Vertices: $(1, 10), \ (1, 0), \ e = \frac{4}{5}$

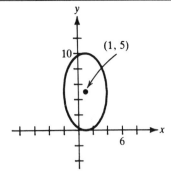

23. Find the center, foci, vertices, and eccentricity of the ellipse $16x^2 + 25y^2 - 32x + 50y + 16 = 0$ and sketch its graph.

Solution:

$$16x^2 + 25y^2 - 32x + 50y + 16 = 0$$
$$16(x^2 - 2x + 1) + 25(y^2 + 2y + 1) = -16 + 16 + 25$$
$$16(x - 1)^2 + 25(y + 1)^2 = 25$$
$$\frac{(x - 1)^2}{25/16} + \frac{(y + 1)^2}{1} = 1$$

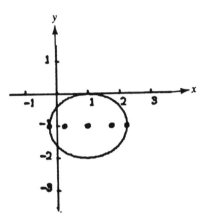

$a^2 = \dfrac{25}{16}, \ b^2 = 1, \ c^2 = \dfrac{9}{16}$

Center: $(1, -1)$
Foci: $\left(\frac{1}{4}, -1\right), \ \left(\frac{7}{4}, -1\right)$
Vertices: $\left(-\frac{1}{4}, -1\right), \ \left(\frac{9}{4}, -1\right), \ e = \frac{3}{5}$

27. Find an equation of the specified ellipse.

Vertices: $(\pm 6, \ 0)$
Foci: $(\pm 5, \ 0)$

Solution:
The major axis is horizontal with the center at $(0, 0)$.
$a = 6, \ c = 5$ implies $b = \sqrt{11}$.

$$\frac{(x - 0)^2}{(6)^2} + \frac{(y - 0)^2}{(\sqrt{11})^2} = 1$$
$$\frac{x^2}{36} + \frac{y^2}{11} = 1$$

31. Find an equation of the specified ellipse.

Vertices: $(0, \ \pm 2)$
Minor axis length 2

Solution:
The major axis is vertical with center at $(0, 0)$.
$a = 2$, $2b = 2 \Rightarrow b = 1$

$$\frac{(x-0)^2}{(1)^2} + \frac{(y-0)^2}{(2)^2} = 1$$

$$x^2 + \frac{y^2}{4} = 1$$

35. Find an equation of the specified ellipse.

> Foci: $(0, 0)$, $(0, 8)$
> Major axis of length 16

Solution:
The major axis is vertical with center at $(0, 4)$.
$2a = 16 \Rightarrow a = 8$

$$c = 4 \Rightarrow b = \sqrt{48} = 4\sqrt{3}$$

$$\frac{(x-0)^2}{(\sqrt{48})^2} + \frac{(y-4)^2}{(8)^2} = 1$$

$$\frac{x^2}{48} + \frac{(y-4)^2}{64} = 1$$

39. Find an equation of the specified ellipse.

> Center: $(3, 2)$, $a = 3c$
> Foci: $(1, 2)$, $(5, 2)$

Solution:
The major axis is horizontal with center at $(3, 2)$.
$c = 2 \Rightarrow a = 3(2) = 6 \Rightarrow b = \sqrt{32}$

$$\frac{(x-3)^2}{(6)^2} + \frac{(y-2)^2}{(\sqrt{32})^2} = 1$$

$$\frac{(x-3)^2}{36} + \frac{(y-2)^2}{32} = 1$$

43. A fireplace arch is to be constructed in the shape of a semiellipse. The opening is to have a height of 2 feet at the center and a width of 5 feet along the base, as shown in the figure. The contractor will first draw the form of the ellipse by the method shown in Figure 11.12. Where should the tacks be placed and how long should the piece of string be?

Solution:

$a = \frac{5}{2}$, $b = 2$, $c = \sqrt{\left(\frac{5}{2}\right)^2 - (2)^2} = \frac{3}{2}$

The tacks should be placed 1.5 feet from the center.

$d_1 + d_2 = 2a = 2\left(\frac{5}{2}\right) = 5$

The string should be 5 feet long.

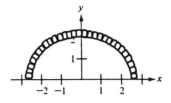

47. The earth moves in an elliptical orbit with the sun at one of the foci. The length of half the major axis is 93 million miles and the eccentricity is 0.017. Find the least and greatest distances between the earth and the sun.

Solution:

$e = \dfrac{c}{a}$, $a = 93,000,000$, $e = 0.017$, $0.017 = \dfrac{c}{93,000,000}$, $c \approx 1,581,000$

Least distance: $a - c \approx 91,419,000$ miles
Greatest distance: $a + c \approx 94,581,000$ miles

51. Show that the equation of an ellipse can be written

$$\frac{(x-h)^2}{a^2} + \frac{(y-k)^2}{a^2(1-e^2)} = 1.$$

Note that as e approaches zero, with a remaining fixed, the ellipse approaches a circle of radius a.

Solution:

$$\frac{x^2}{a^2} + \frac{y^2}{b^2} = 1$$

$$\frac{x^2}{a^2} + \frac{y^2}{a^2(b^2/a^2)} = 1$$

$$\frac{x^2}{a^2} + \frac{y^2}{a^2(a^2 - c^2)/a^2} = 1$$

$$\frac{x^2}{a^2} + \frac{y^2}{a^2(1-e^2)} = 1$$

As $e \Rightarrow 0$, $1 - e^2 \Rightarrow 1$ and we have $\dfrac{x^2}{a^2} + \dfrac{y^2}{a^2} = 1$ or the circle $x^2 + y^2 = a^2$.

SECTION 7.5

Hyperbolas

- A *hyperbola* is the set of all points $(x,\ y)$ the difference of whose distances from two fixed points (*foci*) is constant.

- The standard equation of a hyperbola with center $(h,\ k)$ and transverse and conjugate axes of lengths $2a$ and $2b$ is:

  (a) $\dfrac{(x-h)^2}{a^2} - \dfrac{(y-k)^2}{b^2} = 1$ if the transverse axis is horizontal.

  (b) $\dfrac{(y-k)^2}{a^2} - \dfrac{(x-h)^2}{b^2} = 1$ if the transverse axis is vertical.

- $c^2 = a^2 + b^2$ where c is the distance from the center to a focus.

- The asymptotes of a hyperbola are:

  (a) $y = k \pm \dfrac{b}{a}(x-h)$ if the transverse axis is horizontal.

  (b) $y = k \pm \dfrac{a}{b}(x-h)$ if the transverse axis is vertical.

- The eccentricity of a hyperbola is $e = \dfrac{c}{a}$.

- To classify a nondegenerate conic from its general equation $Ax^2 + Cy^2 + Dx + Ey + F = 0$:
  (a) If $A = C$ $(A \neq 0,\ C \neq 0)$, then it is a circle.
  (b) If $AC = 0$ $(A = 0$ or $C = 0$, but not both$)$, then it is a parabola.
  (c) If $AC > 0$, then it is an ellipse.
  (d) If $AC < 0$, then it is a hyperbola.

Solutions to Selected Exercises

5. Match the following equation with the correct graph.

$$\frac{(x-2)^2}{9} - \frac{y^2}{4} = 1$$

Solution:
$a = 3,\ b = 2$
Center: $(2,\ 0)$
Horizontal transverse axis
Matches graph (d)

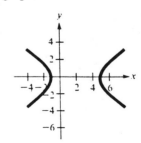

9. Find the center, vertices, and foci of the following hyperbola and sketch its graph, using asymptotes as an aid.

$$\frac{y^2}{1} - \frac{x^2}{4} = 1$$

Solution:
$a = 1,\ b = 2,\ c = \sqrt{5}$
Center: $(0, 0)$
Vertices: $(0, \pm 1)$
Foci: $(0, \pm\sqrt{5})$
Asymptotes: $y = \pm\frac{1}{2}x$

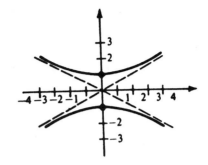

13. Find the center, vertices, and foci of the hyperbola $2x^2 - 3y^2 = 6$ and sketch its graph, using asymptotes as an aid.

Solution:

$$2x^2 - 3y^2 = 6$$

$$\frac{x^2}{3} - \frac{y^2}{2} = 1$$

$a = \sqrt{3},\ b = \sqrt{2},\ c = \sqrt{5}$
Center: $(0, 0)$
Vertices: $(\pm\sqrt{3},\ 0)$
Foci: $(\pm\sqrt{5},\ 0)$
Asymptotes: $y = \pm\sqrt{\frac{2}{3}}\,x$

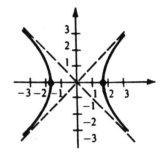

19. Find the center, vertices, and foci of the hyperbola $(y+6)^2 - (x-2)^2 = 1$ and sketch its graph, using asymptotes as an aid.

Solution:

$$\frac{(y+6)^2}{1} - \frac{(x-2)^2}{1} = 1$$

$a = 1,\ b = 1,\ c = \sqrt{2}$
Center: $(2, -6)$
Vertices: $(2, -5),\ (2, -7)$
Foci: $(2, -6 \pm \sqrt{2})$
Asymptotes: $y = -6 \pm (x - 2)$

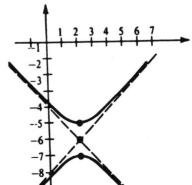

23. Find the center, vertices, and foci of the hyperbola $9y^2 - x^2 + 2x + 54y + 62 = 0$.

Solution:

$$9y^2 - x^2 + 2x + 54y + 62 = 0$$

$$9(y^2 + 6y + 9) - (x^2 - 2x + 1) = -62 - 1 + 81$$

$$\frac{(y + 3)^2}{2} - \frac{(x - 1)^2}{18} = 1$$

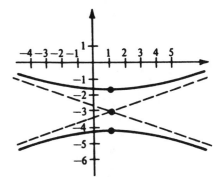

$a = \sqrt{2}$, $b = 3\sqrt{2}$, $c = 2\sqrt{5}$

Center: $(1, -3)$

Vertices: $(1, -3 \pm \sqrt{2})$

Foci: $(1, -3 \pm 2\sqrt{5})$

Asymptotes: $y = -3 \pm \frac{1}{3}(x - 1)$

29. Find an equation of the hyperbola with vertices at $(\pm 1, 0)$, and asymptotes $y = \pm 3x$.

Solution:

$a = 1$, $b/a = 3 \Rightarrow b = 3$

Center: $(0, 0)$

The transverse axis is horizontal.

$$\frac{(x - 0)^2}{(1)^2} - \frac{(y - 0)^2}{(3)^2} = 1$$

$$\frac{x^2}{1} - \frac{y^2}{9} = 1$$

33. Find an equation of the hyperbola with vertices at $(4, 1)$, $(4, 9)$ and foci at $(4, 0)$, $(4, 10)$.

Solution:

$a = 4$, $c = 5 \Rightarrow b = 3$

Center: $(4, 5)$

The transverse axis is vertical.

$$\frac{(y - 5)^2}{(4)^2} - \frac{(x - 4)^2}{(3)^2} = 1$$

$$\frac{(y - 5)^2}{16} - \frac{(x - 4)^2}{9} = 1$$

37. Find an equation of the hyperbola with vertices at $(-2, 1)$, $(2, 1)$ and solution point at $(4, 3)$.

Solution:

$a = 2$

Center: $(0, 1)$

The transverse axis is horizontal.

$$\frac{(x-0)^2}{(2)^2} - \frac{(y-1)^2}{b^2} = 1$$

$$\frac{x^2}{4} - \frac{(y-1)^2}{b^2} = 1$$

$$\frac{(4)^2}{4} - \frac{(3-1)^2}{b^2} = 1$$

$$4 - \frac{4}{b^2} = 1$$

$$4 = 3b^2$$

$$b^2 = \frac{4}{3}$$

$$\frac{x^2}{4} - \frac{(y-1)^2}{4/3} = 1$$

$$\frac{x^2}{4} - \frac{3(y-1)^2}{4} = 1$$

41. Three listening stations located at $(4400, 0)$, $(4400, 1100)$, and $(-4400, 0)$ hear an explosion. If the latter two stations heard the sound 1 second and 5 seconds after the first, respectively, where did the explosion occur? Assume that the coordinate system is measured in feet and that sound travels at the rate of 1100 feet per second.

Solution:

The listening stations are located at $A(4400, 0)$, $B(4400, 1100)$, and $C(-4400, 0)$. Since B heard the explosion 1 second after A heard it, the explosion must have occurred at a point that is 1100 feet farther from B than from A. Since A and B are 1100 feet apart, we conclude that the explosion occurred below A on the line $x = 4400$. Furthermore, since C heard the explosion 5 seconds after A, the explosion must have occurred 5500 feet farther away from C than from A. The collection of all such points is a hyperbola with foci at A and C, center at $(0, 0)$ and $2a = 5500$. Thus,

$$2a = 5500 \quad \text{or} \quad a = 2750$$

$$2c = 8800 \quad \text{or} \quad c = 4400$$

$$b = \sqrt{(4400)^2 - (2750)^2} = 550\sqrt{39}$$

and we have

$$\frac{x^2}{(2750)^2} - \frac{y^2}{(550\sqrt{39})^2} = 1.$$

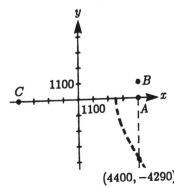

$(4400, -4290)$

Finally, if $x = 4400$ and $y < 0$, we have

$$\frac{y^2}{(550\sqrt{39})^2} = \frac{(4400)^2}{(2750)^2} - 1$$

$$y = -550\sqrt{39}\sqrt{\frac{64-25}{25}} = -110\sqrt{39}\sqrt{39} = -4290$$

45. Classify the graph of the equation $4x^2 - y^2 - 4x - 3 = 0$ as a circle, a parabola, an ellipse, or a hyperbola.

Solution:
$4x^2 - y^2 - 4x - 3 = 0$
$A = 4, \ C = -1$
$AC = (4)(-1) = -4 < 0$
Therefore, the graph is a hyperbola.

49. Classify the graph of the equation $25x^2 - 10x - 200y - 119 = 0$ as a circle, a parabola, an ellipse, or a hyperbola.

Solution:
$25x^2 - 10x - 200y - 119 = 0$
$A = 25, \ C = 0$
$AC = (25)(0) = 0$
Therefore, the graph is a parabola.

SECTION 7.6

Rotation and the General Second-Degree Equation

■ The general second-degree equation $Ax^2 + Bxy + Cy^2 + Dx + Ey + F = 0$ can be rewritten as $A'(x')^2 + C'(y')^2 + D'x' + E'y' + F' = 0$ by rotating the coordinate axes through the angle θ, where $\cot 2\theta = (A - C)/B$.

■ The coefficients of the new system are:

$A' = A\cos^2\theta + B\cos\theta\sin\theta + C\sin^2\theta$
$C' = A\sin^2\theta - B\cos\theta\sin\theta + C\cos^2\theta$
$D' = D\cos\theta + E\sin\theta$
$E' = -D\sin\theta + E\cos\theta$.
$F' = F$

■ The graph of the nondegenerate equation $Ax^2 + Bxy + Cy^2 + Dx + Ey + F = 0$ is:

(a) An ellipse or circle if $B^2 - 4AC < 0$.
(b) A parabola if $B^2 - 4AC = 0$.
(c) A hyperbola if $B^2 - 4AC > 0$.

Solutions to Selected Exercises

3. Rotate the axes to eliminate the xy-term in the equation $9x^2 + 24xy + 16y^2 + 90x - 130y = 0$. Sketch the graph of the resulting equation, showing both sets of axes.

Solution:
$A = 9$, $B = 24$, $C = 16$, $D = 90$, $E = -130$, $F = 0$

$$\cot 2\theta = \frac{A - C}{B} = -\frac{7}{24} \text{ or } \theta = \frac{1}{2}\operatorname{arccot}\left(-\frac{7}{24}\right) \approx 53.13°$$

Since $\cot 2\theta = -\dfrac{7}{24}$, $\cos 2\theta = -\dfrac{7}{25}$.

$$\sin\theta = \frac{\sqrt{1 + (7/25)}}{\sqrt{2}} = \frac{4}{5}$$

$$\cos\theta = \frac{\sqrt{1 - (7/25)}}{\sqrt{2}} = \frac{3}{5}$$

$$A' = 9\left(\frac{9}{25}\right) + 24\left(\frac{4}{5}\right)\left(\frac{3}{5}\right) + 16\left(\frac{16}{25}\right) = 25$$

$$C' = 9\left(\frac{16}{25}\right) - 24\left(\frac{4}{5}\right)\left(\frac{3}{5}\right) + 16\left(\frac{9}{25}\right) = 0$$

$$D' = 90\left(\frac{3}{5}\right) - 130\left(\frac{4}{5}\right) = -50$$

$$E' = -90\left(\frac{4}{5}\right) - 130\left(\frac{3}{5}\right) = -150$$

$$F' = 0$$

$$25x'^2 - 50x' - 150y' = 0$$

$$y' = \frac{x'^2}{6} - \frac{x'}{3}$$

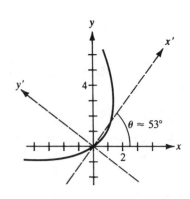

7. Rotate the axes to eliminate the xy-term in the equation $xy - 2y - 4x = 0$. Sketch the graph of the resulting equation, showing both sets of axes.

Solution:

$A = 0,\ B = 1,\ C = 0,\ D = -4,\ E = -2,\ F = 0$

$$\cot 2\theta = 0,\ 2\theta = \frac{\pi}{2},\ \theta = \frac{\pi}{4}$$

$$\sin\theta = \cos\theta = \frac{\sqrt{2}}{2}$$

$$A' = 0 + \frac{1}{2} + 0 = \frac{1}{2}$$

$$C' = 0 - \frac{1}{2} + 0 = -\frac{1}{2}$$

$$D' = -4\left(\frac{\sqrt{2}}{2}\right) - 2\left(\frac{\sqrt{2}}{2}\right) = -3\sqrt{2}$$

$$E' = 4\left(\frac{\sqrt{2}}{2}\right) - 2\left(\frac{\sqrt{2}}{2}\right) = \sqrt{2}$$

$$F' = 0$$

$$\frac{1}{2}x'^2 - \frac{1}{2}y'^2 - 3\sqrt{2}\,x' + \sqrt{2}\,y' = 0$$

$$\frac{(x' - 3\sqrt{2})^2}{16} - \frac{(y' - \sqrt{2})^2}{16} = 1$$

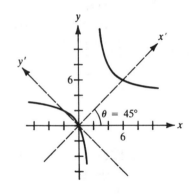

11. Rotate the axes to eliminate the xy-term in the equation $3x^2 - 2\sqrt{3}xy + y^2 + 2x + 2\sqrt{3}y = 0$. Sketch the graph of the resulting equation, showing both sets of axes.

Solution:

$A = 3,\ B = -2\sqrt{3},\ C = 1,\ D = 2,\ E = 2\sqrt{3},\ F = 0$

$$\cot 2\theta = -\frac{1}{\sqrt{3}},\ \theta = 60°$$

$$\sin\theta = \frac{\sqrt{3}}{2},\ \cos\theta = \frac{1}{2}$$

$$A' = 3\left(\frac{1}{4}\right) - 2\sqrt{3}\left(\frac{\sqrt{3}}{2}\right)\left(\frac{1}{2}\right) + \left(\frac{3}{4}\right) = 0$$

$$C' = 3\left(\frac{3}{4}\right) + 2\sqrt{3}\left(\frac{\sqrt{3}}{2}\right)\left(\frac{1}{2}\right) + \left(\frac{1}{4}\right) = 4$$

$$D' = 2\left(\frac{1}{2}\right) + 2\sqrt{3}\left(\frac{\sqrt{3}}{2}\right) = 4$$

$$E' = -2\left(\frac{\sqrt{3}}{2}\right) + 2\sqrt{3}\left(\frac{1}{2}\right) = 0$$

$$F' = 0$$

$$4y'^2 + 4x' = 0$$
$$x' = -(y')^2$$

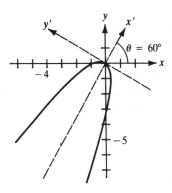

15. Rotate the axes to eliminate the xy-term in the equation $32x^2 + 50xy + 7y^2 = 52$. Sketch the graph of the resulting equation, showing both sets of axes.

Solution:

$A = 32,\ B = 50,\ C = 7,\ D = 0,\ E = 0,\ F = -52$

$\cot 2\theta = \dfrac{1}{2},\ \cos 2\theta = \dfrac{1}{\sqrt{5}},\ \theta \approx 31.72°$

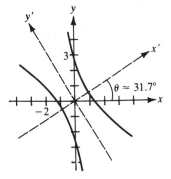

$$\sin\theta = \frac{\sqrt{1 - (1/\sqrt{5})}}{\sqrt{2}} = \sqrt{\frac{\sqrt{5} - 1}{2\sqrt{5}}}$$

$$\cos\theta = \frac{\sqrt{1 + (1/\sqrt{5})}}{\sqrt{2}} = \sqrt{\frac{\sqrt{5} + 1}{2\sqrt{5}}}$$

$$A' = 32\left(\frac{\sqrt{5} + 1}{2\sqrt{5}}\right) + 50\sqrt{\frac{\sqrt{5} - 1}{2\sqrt{5}}}\left(\sqrt{\frac{\sqrt{5} + 1}{2\sqrt{5}}}\right) + 7\left(\frac{\sqrt{5} - 1}{2\sqrt{5}}\right) = \frac{39 + 25\sqrt{5}}{2} \approx 47.451$$

$$C' = 32\left(\frac{\sqrt{5} - 1}{2\sqrt{5}}\right) - 50\sqrt{\frac{\sqrt{5} - 1}{2\sqrt{5}}}\left(\sqrt{\frac{\sqrt{5} + 1}{2\sqrt{5}}}\right) + 7\left(\frac{\sqrt{5} + 1}{2\sqrt{5}}\right) = \frac{39 - 25\sqrt{5}}{2} \approx -8.451$$

$$D' = E' = 0$$

$$F' = -52$$

$$47.451x'^2 - 8.451y'^2 - 52 = 0$$

$$\frac{x'^2}{1.096} - \frac{y'^2}{6.153} = 1$$

19. Use the discriminant to determine whether the graph of $13x^2 - 8xy + 7y^2 - 45 = 0$ is a parabola, an ellipse, or a hyperbola.

Solution:
$A = 13$, $B = -8$, $C = 7$
$B^2 - 4AC = (-8)^2 - 4(13)(7) = -300 < 0$
Therefore, the graph is an ellipse or a circle.

23. Use the discriminant to determine whether the graph of $x^2 + 4xy + 4y^2 - 5x - y - 3 = 0$ is a parabola, an ellipse, or a hyperbola.

Solution:
$A = 1$, $B = 4$, $C = 4$
$B^2 - 4AC = (4)^2 - 4(1)(4) = 0$
Therefore, the graph is a parabola.

25. Show that the equation $x^2 + y^2 = r^2$ is invariant under rotation of axes.

Solution:

$$(x')^2 + (y')^2 = [x \cos \theta + y \sin \theta]^2 + [y \cos \theta - x \sin \theta]^2$$
$$= x^2 \cos^2 \theta + 2xy \cos \theta \sin \theta + y^2 \sin^2 \theta + y^2 \cos^2 \theta - 2xy \cos \theta \sin \theta + x^2 \sin^2 \theta$$
$$= x^2(\cos^2 \theta + \sin^2 \theta) + y^2(\sin^2 \theta + \cos^2 \theta)$$
$$= x^2 + y^2 = r^2$$

REVIEW EXERCISES FOR CHAPTER 7

3. Find an equation of the line through the points $(2, 1)$, $(14, 6)$.

Solution:

$$m = \frac{6-1}{14-2} = \frac{5}{12}$$
$$y - 1 = \frac{5}{12}(x - 2)$$
$$12(y - 1) = 5(x - 2)$$
$$12y - 12 = 5x - 10$$
$$0 = 5x - 12y + 2$$

7. Find t so that the points $(-2, 5)$, $(0, t)$ and $(1, 1)$ are collinear.

Solution:
The line through $(-2, 5)$ and $(1, 1)$ is

$$y - 5 = \frac{1-5}{1+2}(x + 2)$$
$$y - 5 = -\frac{4}{3}(x + 2)$$
$$3y - 15 = -4x - 8$$
$$4x + 3y = 7$$

For $(0, t)$ to be on this line also, it must satisfy the equation $4x + 3y = 7$.

$$4(0) + 3(t) = 7$$

Thus, $t = \frac{7}{3}$.

11. Show that the points $(1, 1)$, $(8, 2)$, $(9, 5)$, and $(2, 4)$ form the vertices of a parallelogram.

Solution:

$$d_1 = \sqrt{(2-1)^2 + (4-1)^2} = \sqrt{10}$$
$$d_2 = \sqrt{(9-2)^2 + (5-4)^2} = \sqrt{50} = 5\sqrt{2}$$
$$d_3 = \sqrt{(8-9)^2 + (2-5)^2} = \sqrt{10}$$
$$d_4 = \sqrt{(1-8)^2 + (1-2)^2} = \sqrt{50} = 5\sqrt{2}$$

Since $d_1 = d_3$ and $d_2 = d_4$, these points are the vertices of a parallelogram.

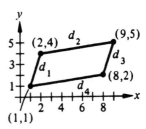

21. Match the following equation with its graph.

$$\frac{x^2}{1} - \frac{y^2}{4} = 1$$

Solution:
This is the standard equation of a hyperbola with vertices $(\pm 1, 0)$ and thus matches the graph shown in (e).

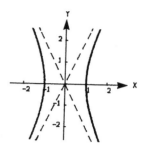

27. Identify and sketch the graph of the rectangular equation $3x^2 + 2y^2 - 12x + 12y + 29 = 0$.

Solution:
Since $AC = 3(2) = 6 > 0$, the graph is an ellipse.

$$3x^2 + 2y^2 - 12x + 12y + 29 = 0$$
$$3(x^2 - 4x + 4) + 2(y^2 + 6y + 9) = -29 + 12 + 18$$
$$3(x - 2)^2 + 2(y + 3)^2 = 1$$
$$\frac{(x - 2)^2}{1/3} + \frac{(y + 3)^2}{1/2} = 1$$

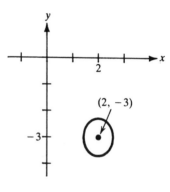

Center: $(2, -3)$

Vertices: $\left(2, -3 \pm \dfrac{\sqrt{2}}{2}\right)$

31. Identify and sketch the graph of the rectangular equation $x^2 + y^2 + 2xy + 2\sqrt{2}\,x - 2\sqrt{2}\,y + 2 = 0$.

Solution:
Since $B^2 - 4AC = 0$, the graph is a parabola.

$A = 1, \ B = 2, \ C = 1, \ D = 2\sqrt{2}, \ E = -2\sqrt{2}, \ F = 2$

$\cot 2\theta = 0, \ 2\theta = \dfrac{\pi}{2}, \ \theta = \dfrac{\pi}{4}, \ \sin \theta = \dfrac{\sqrt{2}}{2}, \ \cos \theta = \dfrac{\sqrt{2}}{2}$

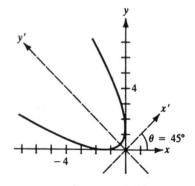

$$A' = \left(\frac{\sqrt{2}}{2}\right)^2 + 2\left(\frac{\sqrt{2}}{2}\right)^2 + \left(\frac{\sqrt{2}}{2}\right)^2 = 2$$
$$C' = \left(\frac{\sqrt{2}}{2}\right)^2 - 2\left(\frac{\sqrt{2}}{2}\right)^2 + \left(\frac{\sqrt{2}}{2}\right)^2 = 0$$
$$D' = 2\sqrt{2}\left(\frac{\sqrt{2}}{2}\right) - 2\sqrt{2}\left(\frac{\sqrt{2}}{2}\right) = 0$$
$$E' = -2\sqrt{2}\left(\frac{\sqrt{2}}{2}\right) - 2\sqrt{2}\left(\frac{\sqrt{2}}{2}\right) = -4$$
$$F' = 2$$

$$2x'^2 - 4y' + 2 = 0$$

$$x'^2 = 4\left(\frac{1}{2}\right)\left(y' - \frac{1}{2}\right)$$

Vertex: $(x', y') = \left(0, \frac{1}{2}\right)$ or $(x, y) = \left(\frac{-1}{\sqrt{8}}, \frac{1}{\sqrt{8}}\right)$

35. Find a rectangular equation for the parabola with vertex at $(0, 2)$ and directrix $x = -3$.

Solution:

$p = 3$, $(h, k) = (0, 2)$

$$(y - k)^2 = 4p(x - h)$$
$$(y - 2)^2 = 4(3)(x - 0)$$
$$y^2 - 4y + 4 = 12x$$
$$y^2 - 12x - 4y + 4 = 0$$

39. Find a rectangular equation for the ellipse with vertices at $(0, \pm 6)$ and passes through the points $(2, 2)$.

Solution:

$(h, k) = (0, 0)$

Vertical major axis with $a = 6$

$$\frac{x^2}{b^2} + \frac{y^2}{36} = 1$$

Since the graph passes through the points $(2, 2)$, we have

$$\frac{4}{b^2} + \frac{4}{36} = 1$$
$$36 + b^2 = 9b^2$$
$$36 = 8b^2$$
$$\frac{36}{8} = b^2$$
$$\frac{9}{2} = b^2$$
$$\frac{x^2}{9/2} + \frac{y^2}{36} = 1$$
$$\frac{2x^2}{9} + \frac{y^2}{36} = 1$$

43. Find a rectangular equation of the hyperbola with foci at $(0, 0)$, $(8, 0)$ and asymptotes $y = \pm 2(x - 4)$.

Solution:

$(h,\ k) = (4,\ 0)$

The transverse axis is horizontal with $c = 4$. Also from the slopes of the asymptotes $\pm b/a = \pm 2$ or $\pm b = \pm 2a$. Now $c^2 = a^2 + b^2 = a^2 + (2a)^2 = 16$. Therefore,

$$a^2 = \frac{16}{3},\ b^2 = \frac{64}{5} \text{ and } \frac{(x-4)^2}{16/5} - \frac{y^2}{64/5} = 1,\ \frac{5(x-4)^2}{16} - \frac{5y^2}{64} = 1.$$

47. Find an equation of the tangent line to

$$\frac{x^2}{100} + \frac{y^2}{25} = 1$$

at the point $(-8,\ 3)$. The tangent line to the conic

$$\frac{x^2}{a^2} \pm \frac{y^2}{b^2} = 1$$

at the point $(x_0,\ y_0)$ is given by

$$\frac{x_0 x}{a^2} \pm \frac{y_0 y}{b^2} = 1.$$

Solution:

$(x_0,\ y_0) = (-8,\ 3)$
$a^2 = 100,\ b^2 = 25$

$$\frac{-8x}{100} + \frac{3y}{25} = 1$$

$$-8x + 12y = 100$$

$$2x - 3y = -25$$

51. Find the points of intersection of the graphs of the equations.

$$x + 3y = 15$$
$$x^2 + y^2 = 25$$

Solution:

$$x + 3y = 15 \Rightarrow x = 15 - 3y$$
$$x^2 + y^2 = 25$$
$$(15 - 3y)^2 + y^2 = 25$$
$$225 - 90y + 9y^2 + y^2 = 25$$
$$10y^2 - 90y + 200 = 0$$
$$y^2 - 9y + 20 = 0$$
$$(y - 4)(y - 5) = 0$$

$y = 4$ $\qquad$ OR $\quad$ $y = 5$

$x = 15 - 3(4) = 3$ $\qquad$ $x = 15 - 3(5) = 0$

The points of intersection are $(3,\ 4)$ and $(0,\ 5)$.

Practice Test for Chapter 7

1. Find the slope of the line passing through the points $(4, 3)$ and $(-2, 7)$.

2. Find the slope and y-intercept of the line $3x - 10y + 40 = 0$.

3. Find an equation for the line passing through the points $(-8, 6)$ and $(-7, 15)$.

4. Find an equation of the line passing through $(-1, 5)$ with slope $m = -\frac{1}{3}$.

5. Find an equation of the line with x-intercept $(6, 0)$ and y-intercept $(0, -2)$.

6. Sue earns a monthly salary of $1950 plus a commission of $5\frac{1}{2}\%$ of her sales. Write a linear equation for her monthly wage W, in terms of her monthly sales, S.

7. Write an equation of the line through the point $(4, -7)$ parallel to the line $x - 4y + 8 = 0$.

8. Write an equation of the line through the point $(11, 8)$ perpendicular to the line $x = 6$.

9. Determine whether the points $(0, 2)$, $(-1, -1)$, and $(5, 15)$ are collinear.

10. Find the distance between the point $(9, 4)$ and the line $2x - 4y + 7 = 0$.

11. Find the inclination of the line $3x + 9y = 20$.

12. Find the angle between the lines $y = 3x - 4$ and $y = -2x + 10$.

13. Find the vertex, focus, and directrix of the parabola $x^2 - 6x - 4y + 1 = 0$.

14. Find an equation of the parabola with its vertex at $(2, -5)$ and focus at $(2, -6)$.

15. Find the center, foci, vertices, and eccentricity of the ellipse $x^2 + 4y^2 - 2x + 32y + 61 = 0$.

16. Find an equation of the ellipse with vertices $(0, \pm 6)$ and eccentricity $e = \frac{1}{2}$.

17. Find the center, vertices, foci, and asymptotes of the hyperbola $16y^2 - x^2 - 6x - 128y + 231 = 0$.

18. Find an equation of the hyperbola with vertices at $(\pm 3, 2)$ and foci at $(\pm 5, 2)$.

19. Rotate the axes to eliminate the xy-term. Sketch the graph of the resulting equation, showing both sets of axes. $5x^2 + 2xy + 5y^2 - 10 = 0$

20. Use the discriminant to determine whether the graph of the equation is a parabola, ellipse, or hyperbola.
    (a) $6x^2 - 2xy + y^2 = 0$ (b) $x^2 + 4xy + 4y^2 - x - y + 17 = 0$

CHAPTER 8

Polar Coordinates and Parametric Equations

SECTION 8.1

Polar Coordinates

- In polar coordinates you do not have unique representation of points. The point $(r,\ \theta)$ can be represented by $(r,\ \theta + 2n\pi)$ or by $(-r,\ \theta + (2n+1)\pi)$ where n is any integer. The pole is represented by $(0,\ \theta)$ where θ is any angle.

- To convert from polar coordinates to rectangular coordinates, use the following relationships.

  $$x = r \cos \theta$$
  $$y = r \sin \theta$$

- To convert from rectangular coordinates to polar coordinates, use the following relationships.

  $$r = \pm\sqrt{x^2 + y^2}$$
  $$\tan \theta = y/x$$

  If θ is in the same quadrant as the point $(x,\ y)$, then r is positive. If θ is in the opposite quadrant as the point $(x,\ y)$, then r is negative.

- You should be able to convert rectangular equations to polar form and vice versa.

Solutions to Selected Exercises

3. Plot the polar point $(-1,\ 5\pi/4)$ and find the corresponding rectangular coordinates.

Solution:

$r = -1,\ \theta = \dfrac{5\pi}{4}$

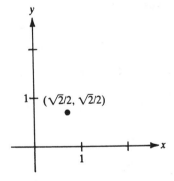

$$x = (-1)\cos\frac{5\pi}{4} = (-1)\left(-\frac{\sqrt{2}}{2}\right) = \frac{\sqrt{2}}{2}$$

$$y = (-1)\sin\frac{5\pi}{4} = (-1)\left(-\frac{\sqrt{2}}{2}\right) = \frac{\sqrt{2}}{2}$$

$\left(-1,\ \dfrac{5\pi}{4}\right)$ corresponds to $\left(\dfrac{\sqrt{2}}{2},\ \dfrac{\sqrt{2}}{2}\right)$.

7. Plot the polar point $(\sqrt{2},\ 2.36)$ and find the corresponding rectangular coordinates.

Solution:

$r = \sqrt{2}$, $\theta = 2.36$ (in radians)

$$x = \sqrt{2}\cos 2.36 \approx -1.004$$
$$y = \sqrt{2}\sin 2.36 \approx 0.996$$

$(\sqrt{2},\ 2.36)$ corresponds to $(-1.004,\ 0.996)$.

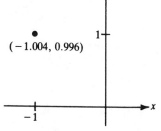

11. Find two sets of polar coordinates for the rectangular point $(-3,\ 4)$, using $0 \le \theta < 2\pi$.

Solution:

$x = -3$, $y = 4$

$r = \pm\sqrt{(-3)^2 + (4)^2} = \pm 5$

$\tan\theta = -\frac{4}{3}$, $\theta \approx 2.214$

The point $(-3,\ 4)$ is in Quadrant II as is the angle $\theta = 2.214$. Thus, one polar representation is $(5,\ 2.214)$. Another representation is $(-5,\ 2.214 + \pi) \approx (-5,\ 5.356)$.

15. Find two sets of polar coordinates for the rectangular point $(4,\ 6)$, using $0 \le \theta < 2\pi$.

Solution:

$x = 4$, $y = 6$

$r = \pm\sqrt{(4)^2 + (6)^2} = \pm 2\sqrt{13}$

$\tan\theta = \frac{6}{4}$, $\theta \approx 0.983$

Since $(4,\ 6)$ is in Quadrant I and $\theta = 0.983$ is in Quadrant I, one representation in polar coordinates is $(2\sqrt{13},\ 0.983)$. Another representation is $(-2\sqrt{13},\ 0.983 + \pi) \approx (-2\sqrt{13},\ 4.124)$.

19. Convert the equation $x^2 + y^2 - 2ax = 0$ to polar form.

Solution:

$$x^2 + y^2 - 2ax = 0$$
$$r^2 - 2ar\cos\theta = 0$$
$$r(r - 2a\cos\theta) = 0$$
$$r = 0 \ \ \text{OR} \ \ r = 2a\cos\theta$$

Since $r = 0$ is the pole and is also on the graph of $r = 2a\cos\theta$, we just have $r = 2a\cos\theta$.

23. Convert the equation $x = 10$ to polar form.

Solution:

$$x = 10$$
$$r\cos\theta = 10$$
$$r = \frac{10}{\cos\theta}$$
$$r = 10\sec\theta$$

27. Convert the equation $xy = 4$ to polar form.

Solution:

$$xy = 4$$
$$(r\cos\theta)(r\sin\theta) = 4$$
$$r^2 = \frac{4}{\cos\theta\sin\theta}$$
$$r^2 = \frac{4}{\frac{1}{2}\sin 2\theta}$$
$$r^2 = 8\csc 2\theta$$

31. Convert the equation $r = 4\sin\theta$ to rectangular form.

Solution:

$$r = 4\sin\theta$$
$$r^2 = 4r\sin\theta$$
$$x^2 + y^2 = 4y$$
$$x^2 + y^2 - 4y = 0$$

35. Convert the equation $r = 2\csc\theta$ to rectangular form.

Solution:

$$r = 2\csc\theta$$
$$r = \frac{2}{\sin\theta}$$
$$r\sin\theta = 2$$
$$y = 2$$

39. Convert the equation $r = \dfrac{6}{2 - 3\sin\theta}$ to rectangular form.

Solution:

$$r = \frac{6}{2 - 3\sin\theta}$$
$$r(2 - 3\sin\theta) = 6$$
$$2r - 3r\sin\theta = 6$$
$$2r = 6 + 3r\sin\theta$$
$$2(\pm\sqrt{x^2 + y^2}) = 6 + 3y$$
$$4(x^2 + y^2) = (6 + 3y)^2$$
$$4x^2 + 4y^2 = 36 + 36y + 9y^2$$
$$4x^2 - 5y^2 - 36y - 36 = 0$$

41. Show that the distance between $(r_1,\ \theta_1)$ and $(r_2,\ \theta_2)$ is $\sqrt{r_1{}^2 + r_2{}^2 - 2r_1r_2\cos(\theta_1 - \theta_2)}$.

Solution:

$(r_1,\ \theta_1)$ corresponds to the point $(r_1\cos\theta_1,\ r_1\sin\theta_1)$ in rectangular coordinates. Likewise, $(r_2,\ \theta_2)$ corresponds to the point $(r_2\cos\theta_2,\ r_2\sin\theta_2)$. In rectangular coordinates we use the distance formula, $d = \sqrt{(x_2 - x_1)^2 + (y_2 - y_1)^2}$, to find the distance between two points.

$$
\begin{aligned}
d &= \sqrt{(r_2\cos\theta_2 - r_1\cos\theta_1)^2 + (r_2\sin\theta_2 - r_1\sin\theta_1)^2} \\
&= \sqrt{r_2{}^2\cos^2\theta_2 - 2r_1r_2\cos\theta_1\cos\theta_2 + r_1{}^2\cos^2\theta_1 + r_2{}^2\sin^2\theta_2 - 2r_1r_2\sin\theta_1\sin\theta_2 + r_1{}^2\sin^2\theta_1} \\
&= \sqrt{r_1{}^2(\cos^2\theta_1 + \sin^2\theta_1) + r_2{}^2(\cos^2\theta_2 + \sin^2\theta_2) - 2r_1r_2(\cos\theta_1\cos\theta_2 + \sin\theta_1\sin\theta_2)} \\
&= \sqrt{r_1{}^2(1) + r_2{}^2(1) - 2r_1r_2\cos(\theta_1 - \theta_2)} \\
&= \sqrt{r_1{}^2 + r_2{}^2 - 2r_1r_2\cos(\theta_1 - \theta_2)}
\end{aligned}
$$

SECTION 8.2

Graphs of Polar Equations

■ When graphing polar equations:

1. Test for symmetry
   (a) $r = f(\sin\theta)$ is symmetric with respect to the line $\theta = \pi/2$.
   (b) $r = f(\cos\theta)$ is symmetric with respect to the polar axis.

2. Find the θ values for which $|r|$ is maximum.

3. Find the θ values for which $r = 0$.

4. Know the different types of polar graphs.
   (a) Limaçons
   $$r = a \pm b\cos\theta$$
   $$r = a \pm b\sin\theta$$
   (b) Rose Curves, $n \geq 2$
   $$r = a\cos n\theta$$
   $$r = a\sin n\theta$$
   (c) Circles
   $$r = a\cos\theta$$
   $$r = a\sin\theta$$
   (d) Lemniscates
   $$r^2 = a^2\cos 2\theta$$
   $$r^2 = a^2\sin 2\theta$$

5. Plot additional points.

Solutions to Selected Exercises

3. Sketch the graph of $\theta = \pi/6$.

Solution:

$$\theta = \frac{\pi}{6}$$

$$\tan\theta = \tan\frac{\pi}{6}$$

$$\frac{y}{x} = \frac{1}{\sqrt{3}}$$

$$\sqrt{3}\,y = x$$

$$y = \frac{x}{\sqrt{3}} \quad \text{Straight line}$$

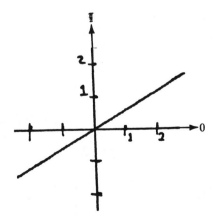

7. Sketch the graph of $r = 3(1 - \cos\theta)$.

Solution:

$a/b = 1$, so the graph is a cardioid.
Symmetric to the polar axis since r is a function of $\cos\theta$.
Maximum value of $|r|$ is 6 and occurs when $\theta = \pi$.
The zero of r occurs when $\theta = 0$.

θ	0	$\dfrac{\pi}{3}$	$\dfrac{\pi}{2}$	$\dfrac{2\pi}{3}$	π
r	0	$\dfrac{3}{2}$	3	$\dfrac{9}{2}$	6

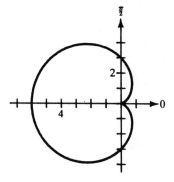

11. Sketch the graph of $r = 2 + 3\sin\theta$.

Solution:

$a/b = 2/3 < 1$, so the graph is a limaçon with inner loop.
Symmetric to $\theta = \pi/2$ since r is a function of $\sin\theta$. Maximum value of $|r|$ is 5 and occurs when $\theta = \pi/2$. The zero of r occurs when $\theta \approx 2.412$.

θ	$-\pi$	$-\dfrac{\pi}{6}$	0	$\dfrac{\pi}{6}$	$\dfrac{\pi}{2}$	π	2.412	$\dfrac{3\pi}{2}$
r	2	$\dfrac{1}{2}$	2	$\dfrac{7}{2}$	5	2	0	-1

15. Sketch the graph of $r = 3 - 2\cos\theta$.

Solution:

$a/b = 3/2 > 1$, so the graph is a dimpled limaçon. Symmetric to the polar axis since r is a function of $\cos\theta$. Maximum value of $|r|$ is 5 and occurs when $\theta = \pi$. There are no zeros of r.

θ	0	$\dfrac{\pi}{3}$	$\dfrac{\pi}{2}$	$\dfrac{2\pi}{3}$	π
r	1	2	3	4	5

19. Sketch the graph of $r = 2\cos 3\theta$.

Solution:

The graph is a rose curve with three petals. Symmetric to the polar axis. Maximum value of $|r|$ is 2 and occurs when $\theta = 0$, $\theta = \pi/3$, and $\theta = 2\pi/3$. Zeros of r occur when $\theta = \pi/6$, $\theta = \pi/2$, and $\theta = 5\pi/6$.

θ	0	$\dfrac{\pi}{2}$	$\dfrac{\pi}{6}$	$\dfrac{\pi}{4}$	$\dfrac{\pi}{3}$	$\dfrac{5\pi}{12}$	$\dfrac{\pi}{2}$
r	2	$\sqrt{2}$	0	$-\sqrt{2}$	-2	$-\sqrt{2}$	0

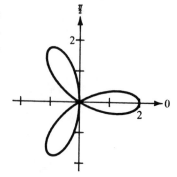

23. Sketch the graph of $r = 2 \sec \theta$.

Solution:

$$r = 2 \sec \theta$$
$$r = \frac{2}{\cos \theta}$$
$$r \cos \theta = 2$$
$$x = 2 \text{ is a vertical line.}$$

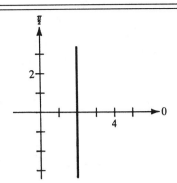

27. Sketch the graph $r^2 = 4 \cos 2\theta$.

Solution:
The graph is a lemniscate. Symmetric to the polar axis, the line $\theta = \pi/2$, and the pole. Maximum value of $|r|$ is 2 and occurs when $\theta = 0$. The zeros of r occur when $\theta = \pi/4$ and $\theta = 3\pi/4$.

θ	0	$\dfrac{\pi}{6}$	$\dfrac{\pi}{4}$
r	± 2	$\pm \sqrt{2}$	0

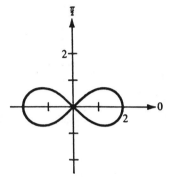

33. Write the equation for the limaçon $r = 2 - \sin \theta$ after it has been rotated by the given amount.

(a) $\dfrac{\pi}{4}$ (b) $\dfrac{\pi}{2}$ (c) π (d) $\dfrac{3\pi}{2}$

Solution:
Refer to Exercises 31 and 32.

(a) $r = 2 - \sin\left(\theta - \dfrac{\pi}{4}\right)$

$\quad = 2 - \dfrac{\sqrt{2}}{2}(\sin \theta + \cos \theta)$

(b) $r = 2 - (-\cos \theta) = 2 + \cos \theta$

(c) $r = 2 - (-\sin \theta) = 2 + \sin \theta$

(d) $r = 2 - \cos \theta$

35. Sketch the graphs of the equations.

(a) $r = 1 - \sin \theta$ (b) $r = 1 - \sin\left(\theta - \dfrac{\pi}{4}\right)$

Solution:

(a) Cardioid

θ	0	$\dfrac{\pi}{2}$	π	$\dfrac{3\pi}{2}$
r	1	0	1	2

(b) Rotate the graph of $r = 1 - \sin\theta$ through the angle $\pi/4$.

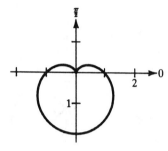

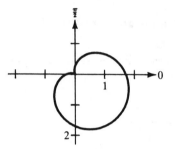

SECTION 8.3

Polar Equations of Conics

- The graph of a polar equation of the form

$$r = \frac{ep}{1 \pm e \cos \theta} \quad \text{or} \quad r = \frac{ep}{1 \pm e \sin \theta}$$

is a conic, where $e > 0$ is the eccentricity and $|p|$ is the distance between the focus (pole) and the directrix.

(a) If $e < 1$, the graph is an ellipse.
(b) If $e = 1$, the graph is a parabola.
(c) If $e > 1$, the graph is a hyperbola.

- Guidelines for finding polar equations of conics:

(a) Horizontal directrix above the pole: $r = \dfrac{ep}{1 + e \sin \theta}$

(b) Horizontal directrix below the pole: $r = \dfrac{ep}{1 - e \sin \theta}$

(c) Vertical directrix to the right of the pole: $r = \dfrac{ep}{1 + e \cos \theta}$

(d) Vertical directrix to the left of the pole: $r = \dfrac{ep}{1 - e \cos \theta}$

Solutions to Selected Exercises

1. Match the following with one of the given graphs.

$$r = \frac{6}{1 - \cos \theta}$$

Solution:
$e = 1$, so the graph is a parabola.
Vertex: $(3, \pi)$
Matches graph (c)

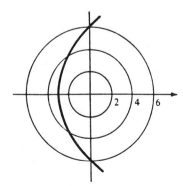

7. Identify and sketch the graph of $r = \dfrac{2}{1 - \cos \theta}$.

Solution:

$e = 1$, so the graph is a parabola.

Vertex: $(1, \pi)$

θ	$\dfrac{\pi}{2}$	π	$\dfrac{3\pi}{2}$
r	2	1	2

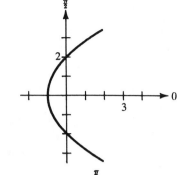

11. Identify and sketch the graph of $r = \dfrac{2}{2 + \cos \theta}$.

Solution:

$$r = \frac{2}{2 + \cos \theta}$$
$$r = \frac{1}{1 + \frac{1}{2} \cos \theta}$$

$e = \frac{1}{2} < 1$, so the graph is an ellipse.

θ	0	$\dfrac{\pi}{2}$	π	$\dfrac{3\pi}{2}$
r	$\dfrac{2}{3}$	1	2	1

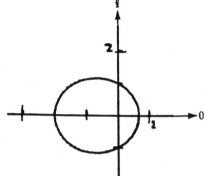

15. Identify and sketch the graph of $r = \dfrac{5}{1 + 2 \cos \theta}$.

Solution:

$e = 2 > 1$, so the graph is a hyperbola.

θ	0	$\dfrac{\pi}{2}$	π	$\dfrac{3\pi}{2}$
r	$\dfrac{5}{3}$	5	-5	5

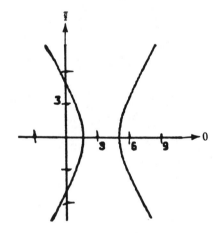

19. Identify and sketch the graph of $r = \dfrac{3}{2 - 6 \cos \theta}$.

Solution:

$$r = \frac{3}{2 - 6 \cos \theta}$$
$$r = \frac{3/2}{1 - 3 \cos \theta}$$

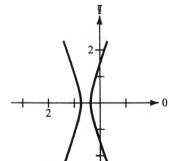

$e = 3 > 1$, so the graph is a hyperbola.

θ	0	$\dfrac{\pi}{2}$	π	$\dfrac{3\pi}{2}$
r	$-\dfrac{3}{4}$	$\dfrac{3}{2}$	$\dfrac{3}{8}$	$\dfrac{3}{2}$

23. Find a polar equation of the ellipse with focus at $(0, 0)$, $e = \frac{1}{2}$, and directrix $y = 1$.

Solution:

$e = \dfrac{1}{2}, \; y = 1, \; p = 1$

$$r = \frac{ep}{1 + e \sin \theta}$$

$$r = \frac{\frac{1}{2}}{1 + \frac{1}{2} \sin \theta}$$

$$r = \frac{1}{2 + \sin \theta}$$

27. Find a polar equation of the parabola with focus at $(0, 0)$ and vertex at $(1, \, -\pi/2)$.

Solution:

$e = 1, \; p = 2$

$$r = \frac{ep}{1 - e \sin \theta}$$

$$r = \frac{2}{1 - \sin \theta}$$

31. Find a polar equation of the hyperbola with focus at $(0, 0)$ and vertices at $(1, \, 3\pi/2)$, $(9, \, 3\pi/2)$.

Solution:

Center: $\left(5, \; \dfrac{3\pi}{2}\right)$; $c = 5, \; a = 4, \; e = \dfrac{c}{a} = \dfrac{5}{4}$

$$r = \frac{ep}{1 - e \sin \theta}$$

$$r = \frac{\frac{5}{4}p}{1 - \frac{5}{4} \sin \theta}$$

$$r = \frac{5p}{4 - 5 \sin \theta}$$

$$1 = \frac{5p}{4 - 5 \sin \frac{3\pi}{2}}$$

$$1 = \frac{5p}{9}$$

$$p = \frac{9}{5}$$

$$r = \frac{5\left(\frac{9}{5}\right)}{4 - 5 \sin \theta}$$

$$r = \frac{9}{4 - 5 \sin \theta}$$

35. Find a polar equation of the parabola with focus at $(0, 0)$ and vertex at $(5, \pi)$.

Solution:
Directrix: $x = -10$, $e = 1$, $p = 10$

$$r = \frac{ep}{1 - e \cos \theta}$$

$$r = \frac{10}{1 - \cos \theta}$$

39. Use the results of Exercises 37 and 38 to write the polar form of

$$\frac{x^2}{169} + \frac{y^2}{144} = 1.$$

Solution:
$a = 13$, $b = 12$, $c = 5$, $e = \frac{5}{13}$

$$r^2 = \frac{b^2}{1 - e^2 \cos^2 \theta}$$

$$r^2 = \frac{144}{1 - \left(\frac{25}{169}\right) \cos^2 \theta}$$

43. Use the results of Exercises 37 and 38 to write the polar form of the hyperbola with one focus at $(5, \pi/2)$ and vertices at $(4, \pi/2)$, $(4, -\pi/2)$.

Solution:
Center: $(0, 0)$, $a = 4$, $c = 5$, $b = 3$, $e = \frac{5}{4}$

$$\frac{x^2}{16} - \frac{y^2}{9} = 1$$

$$r^2 = \frac{-b^2}{1 - e^2 \cos^2 \theta}$$

$$r^2 = \frac{-9}{1 - \left(\frac{25}{16}\right) \cos^2 \theta}$$

47. An earth satellite in a 100-mile-high circular orbit around the earth has a velocity of approximately 17,500 miles per hour. If this velocity is multiplied by $\sqrt{2}$, then the satellite will have the minimum velocity necessary to escape the earth's gravity and it will follow a parabolic path with the center of the earth as the focus. Find a polar equation of the parabolic path of the satellite (assume the radius of the earth is 4000 miles).

Solution:
Directrix: $y = 8200$, $e = 1$, $p = 8200$

$$r = \frac{ep}{1 + e \sin \theta}$$

$$r = \frac{8200}{1 + \sin \theta}$$

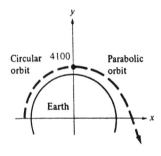

SECTION 8.4

Plane Curves and Parametric Equations

- If f and g are continuous functions of t on an interval I, then the set of ordered pairs $(f(t),\ g(t))$ is called a *plane curve C*. The equations $x = f(t)$ and $y = g(t)$ are called *parametric equations* for C and t is called the *parameter*.

- To eliminate the parameter:
  (a) Solve for t in one equation and substitute into the second equation.
  (b) Use trigonometric identities.

- You should be able to find the parametric equations for a graph.

Solutions to Selected Exercises

5. Sketch the curve represented by the parametric equations, $x = \sqrt{t}$ and $y = 1 - t$, and write the corresponding rectangular equation by eliminating the parameter.

Solution:
Since $x = \sqrt{t}$, we know that $x \geq 0$. $x^2 = t$ substituted into $y = 1 - t$ yields the equation $y = 1 - x^2$, $x \geq 0$.

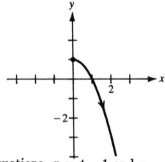

9. Sketch the curve represented by the parametric equations, $x = t - 1$ and $y = t/(t - 1)$, and write the corresponding rectangular equation by eliminating the parameter.

Solution:

$$x = t - 1 \Rightarrow t = x + 1$$
$$y = \frac{t}{t - 1} = \frac{x + 1}{(x + 1) - 1} = \frac{x + 1}{x}$$

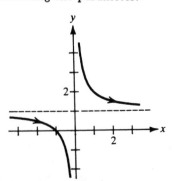

13. Sketch the curve represented by the parametric equations, $x = 4\sin 2\theta$ and $y = 2\cos 2\theta$, and write the corresponding rectangular equation by eliminating the parameter.

Solution:

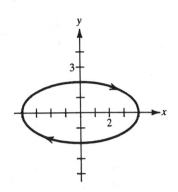

$$x = 4\sin 2\theta \Rightarrow \sin 2\theta = \frac{x}{4}$$

$$y = 2\cos 2\theta \Rightarrow \cos 2\theta = \frac{y}{2}$$

$$\sin^2 2\theta + \cos^2 2\theta = 1$$

$$\left(\frac{x}{4}\right)^2 + \left(\frac{y}{2}\right)^2 = 1$$

$$\frac{x^2}{16} + \frac{y^2}{4} = 1$$

17. Sketch the curve represented by the parametric equations, $x = \sec\theta$ and $y = \cos\theta$, and write the corresponding rectangular equation by eliminating the parameter.

Solution:
Since $x = \sec\theta$, we have the restriction $x \leq -1$ or $x \geq 1$.
Since $y = \cos\theta$, we have the restriction $-1 \leq y \leq 1$.

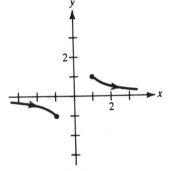

$$x = \sec\theta \Rightarrow \frac{1}{x} = \cos\theta$$

$$y = \cos\theta \Rightarrow y = \frac{1}{x}, \quad x \leq -1 \ \text{OR} \ x \geq 1, \ -1 \leq y \leq 1$$

21. Sketch the curve represented by the parametric equations, $x = 4 + 2\cos\theta$ and $y = -1 + 4\sin\theta$, and write the corresponding rectangular equation by eliminating the parameter.

Solution:

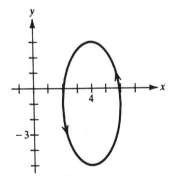

$$x = 4 + 2\cos\theta \Rightarrow \frac{x-4}{2} = \cos\theta$$

$$y = -1 + 4\sin\theta \Rightarrow \frac{y+1}{4} = \sin\theta$$

$$\left(\frac{x-4}{2}\right)^2 + \left(\frac{y+1}{4}\right)^2 = 1$$

$$\frac{(x+4)^2}{4} + \frac{(y+1)^2}{16} = 1$$

25. Sketch the curve represented by the parametric equations, $x = e^{-t}$ and $y = e^{3t}$, and write the corresponding rectangular equation by eliminating the parameter.

Solution:

Since $x = e^{-t}$ and $y = e^{3t}$, we have $x > 0$ and $y > 0$.

$$x = e^{-t} \Rightarrow \frac{1}{x} = e^t$$

$$y = e^{3t} = (e^t)^3 = \left(\frac{1}{x}\right)^3 = \frac{1}{x^3}$$

where $x > 0$ and $y > 0$.

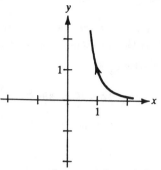

29. Eliminate the parameter and obtain the standard form of the rectangular equation of the curve.

Ellipse: $x = h + a\cos\theta$, $y = k + b\sin\theta$

Solution:

$$x = h + a\cos\theta \Rightarrow \frac{x - h}{a} = \cos\theta$$

$$y = k + b\sin\theta \Rightarrow \frac{y - k}{b} = \sin\theta$$

$$\left(\frac{x - h}{a}\right)^2 + \left(\frac{y - k}{b}\right)^2 = 1$$

$$\frac{(x - h)^2}{a^2} + \frac{(y - k)^2}{b^2} = 1$$

33. Find a set of parametric equations for the circle with center at $(2, 1)$ and radius 4.

Solution:

From Exercise 28 we have $x = h + r\cos\theta$, $y = k + r\sin\theta$. Using $h = 2$, $k = 1$ and $r = 4$, we have $x = 2 + 4\cos\theta$, $y = 1 + 4\sin\theta$. This solution is not unique.

37. Find a set of parametric equations for the hyperbola with vertices at $(\pm 4, 0)$ and foci at $(\pm 5, 0)$.

Solution:

From Exercise 30 we have $x = h + a\sec\theta$, $y = k + b\tan\theta$. Using $(h, k) = (0, 0)$, $a = 4$, $c = 5$, and $b = 3$, we have $x = 4\sec\theta$, $y = 3\tan\theta$. This solution is not unique.

39. Find two different sets of parametric equations for the rectangular equation $y = x^3$.

Solution:

<u>Examples</u>

$$x = t, \qquad y = t^3$$
$$x = \sqrt[3]{t}, \qquad y = t$$
$$x = \tan t, \qquad y = \tan^3 t$$
$$x = t - 4, \qquad y = (t - 4)^3$$

and so on.

43. Sketch the curve represented by the parametric equations.

Witch of Agnesi: $x = 2\cot\theta$, $y = 2\sin^2\theta$

Solution:

$x = 2\cot\theta \Rightarrow \theta = \operatorname{arccot}\dfrac{x}{2}$

$y = 2\sin^2\theta \Rightarrow y = 2\sin^2\left(\operatorname{arccot}\dfrac{x}{2}\right)$

$y = 2\left(\dfrac{2}{\sqrt{x^2+4}}\right)^2$

$y = \dfrac{8}{x^2+4}$

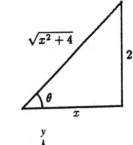

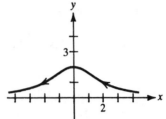

47. A wheel of radius a rolls along a straight line without slipping, as shown in the figure. Find the parametric equation for the curve described by a point P that is b units from the center of the wheel. This curve is called a *curtate cycloid* when $b < a$.

Solution:
When the circle has rolled θ radians, we know that the center is at $(a\theta,\ a)$.

$\sin\theta = \sin(180° - \theta) = \dfrac{|AC|}{b} = \dfrac{|BD|}{b}$ or $|BD| = b\sin\theta$

$\cos\theta = -\cos(180° - \theta) = \dfrac{|AP|}{-b}$ or $|AP| = -b\cos\theta$

Therefore, $x = a\theta - b\sin\theta$ and $y = a - b\cos\theta$.

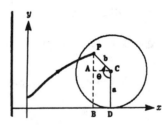

REVIEW EXERCISES FOR CHAPTER 8

1. The polar coordinates of a point are $(-2, 4\pi/3)$. Plot the point and find the rectangular coordinates of the point.

Solution:

$$x = r \cos \theta = -2 \cos \frac{4\pi}{3} = -2\left(-\frac{1}{2}\right) = 1$$

$$y = r \sin \theta = -2 \sin \frac{4\pi}{3} = -2\left(-\frac{\sqrt{3}}{2}\right) = \sqrt{3}$$

The rectangular coordinates are $(1, \sqrt{3})$.

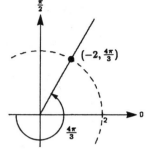

5. The rectangular coordinates of a point are $(0, 8)$. Find two sets of polar coordinates for the point, using $0 \le \theta < 2\pi$.

Solution:

$$r = \pm\sqrt{(0)^2 + (8)^2} = \pm 8$$

$$\tan \theta = \frac{8}{0} \quad \text{is undefined.}$$

$$\theta = \frac{\pi}{2} \quad \text{or} \quad \theta = \frac{3\pi}{2}$$

The polar coordinates are $(8, \pi/2)$ or $(-8, 3\pi/2)$.

9. Identify and sketch the graph of the polar equation $r = 4$.

Solution:
$r = 4$ is a circle centered at the pole of radius 4.

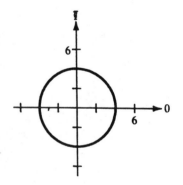

13. Identify and sketch the graph of the polar equation $r = -2(1 + \cos \theta)$.

Solution:
$r = -2(1 + \cos \theta)$ is a cardioid; symmetric to the polar axis.

θ	0	$\dfrac{\pi}{3}$	$\dfrac{\pi}{2}$	$\dfrac{2\pi}{3}$	π
r	-4	-3	-2	-1	0

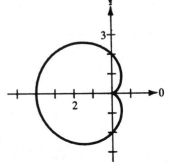

17. Identify and sketch the graph of the polar equation $r = -3\cos 2\theta$.

Solution:

$r = -3\cos 2\theta$ is a rose curve with four petals; symmetric to the polar axis, $\theta = \pi/2$, and the pole. Maximum value of $|r|$ is 3.

$(-3, 0),\ \left(3,\ \dfrac{\pi}{2}\right),\ (-3,\ \pi),\ \left(3,\ \dfrac{3\pi}{2}\right)$

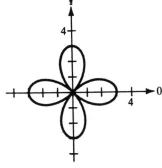

21. Identify and sketch the graph of the polar equation $r = \dfrac{3}{\cos(\theta - (\pi/4))}$.

Solution:

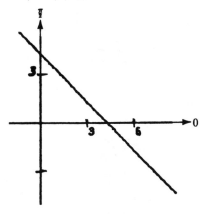

$$r = \frac{3}{\cos(\theta - (\pi/4))}$$

$$r\cos\left(\theta - \frac{\pi}{4}\right) = 3$$

$$r\left[\cos\theta\cos\frac{\pi}{4} + \sin\theta\sin\frac{\pi}{4}\right] = 3$$

$$r\left[\frac{\sqrt{2}}{2}\cos\theta + \frac{\sqrt{2}}{2}\sin\theta\right] = 3$$

$$\frac{\sqrt{2}}{2}r\cos\theta + \frac{\sqrt{2}}{2}r\sin\theta = 3$$

$$r\cos\theta + r\sin\theta = 3\sqrt{2}$$

$$x + y = 3\sqrt{2} \quad \text{Line}$$

25. Convert $r = 3\cos\theta$ to rectangular form.

Solution:

$$r = 3\cos\theta$$

$$r^2 = 3r\cos\theta$$

$$x^2 + y^2 = 3x$$

$$x^2 + y^2 - 3x = 0$$

29. Convert $r^2 = \cos 2\theta$ to rectangular form.

Solution:

$$r^2 = \cos 2\theta$$
$$r^2 = 2\cos^2\theta - 1$$
$$x^2 + y^2 = 2\left(\frac{x^2}{x^2 + y^2}\right) - 1$$
$$(x^2 + y^2)^2 = 2x^2 - (x^2 + y^2)$$
$$x^4 + 2x^2y^2 + y^4 = x^2 - y^2$$
$$x^4 + 2x^2y^2 + y^4 - x^2 + y^2 = 0$$

33. Find a polar equation for a parabola with vertex at $(2,\ \pi)$ and focus at $(0,\ 0)$.

Solution:

$e = 1,\ p = 4$

$$r = \frac{ep}{1 - e\cos\theta}$$
$$r = \frac{4}{1 - \cos\theta}$$

37. Find a polar equation for a circle with center at $(0,\ 5)$ and passes through $(0,\ 0)$.

Solution:

The radius is 5.

$$x^2 + (y - 5)^2 = 25$$
$$x^2 + y^2 - 10y = 0$$
$$r^2 - 10r\sin\theta = 0$$
$$r(r - 10\sin\theta) = 0$$
$$r = 10\sin\theta$$

41. Sketch the curve represented by the parametric equations, $x = 1 + 4t$, $y = 2 - 3t$, and if possible, write the corresponding rectangular equation by eliminating the parameter.

Solution:

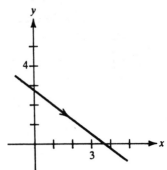

$$x = 1 + 4t \Rightarrow t = \frac{x - 1}{4}$$
$$y = 2 - 3t \Rightarrow y = 2 - 3\left(\frac{x - 1}{4}\right)$$
$$y = -\frac{3}{4}x + \frac{11}{4}$$
$$3x + 4y = 11$$

45. Sketch the curve represented by the parametric equations, $x = 6\cos\theta$, $y = 6\sin\theta$, and if possible, write the corresponding rectangular equation by eliminating the parameter.

Solution:

$$x = 6\cos\theta \Rightarrow \cos\theta = \frac{x}{6}$$

$$y = 6\sin\theta \Rightarrow \sin\theta = \frac{y}{6}$$

$$\left(\frac{x}{6}\right)^2 + \left(\frac{y}{6}\right)^2 = 1$$

$$x^2 + y^2 = 36$$

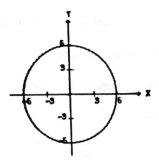

49. Sketch the curve represented by the parametric equations, $x = \cos^3\theta$, $y = 4\sin^3\theta$, and if possible, write the corresponding rectangular equation by eliminating the parameter.

Solution:

$$x = \cos^3\theta \Rightarrow \cos\theta = x^{1/3}$$

$$y = 4\sin^3\theta \Rightarrow \sin\theta = \left(\frac{y}{4}\right)^{1/3}$$

$$(x^{1/3})^2 + \left[\left(\frac{y}{4}\right)^{1/3}\right]^2 = 1$$

$$x^{2/3} + \left(\frac{y}{4}\right)^{2/3} = 1$$

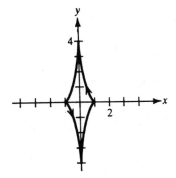

Practice Test for Chapter 8

1. Convert the polar point $\left(\sqrt{2},\ 3\pi/4\right)$ to rectangular coordinates.

2. Convert the rectangular point $\left(\sqrt{3},\ -1\right)$ to polar coordinates.

3. Convert the rectangular equation $4x - 3y = 12$ to polar form.

4. Convert the polar equation $r = 5\cos\theta$ to rectangular form.

5. Convert the polar equation $r = \dfrac{3}{4 + \cos\theta}$ to rectangular form.

6. Sketch the graph of $r = 1 - \cos\theta$.

7. Sketch the graph of $\theta = \dfrac{3\pi}{4}$.

8. Sketch the graph of $r = 4\csc\theta$.

9. Sketch the graph of $r^2 = 9\cos 2\theta$.

10. Sketch the graph of $r = 5\sin 2\theta$.

11. Sketch the graph of $r = \dfrac{12}{6\sin\theta - \cos\theta}$.

12. Sketch the graph of $r = \dfrac{3}{6 - \cos\theta}$.

13. Sketch the graph of $r = \dfrac{5}{4 + 4\sin\theta}$.

14. Sketch the graph of $r = \dfrac{10}{1 - 5\cos\theta}$.

15. Find a polar equation of the parabola with its vertex at $\left(6,\ \pi/2\right)$ and focus at $(0,\ 0)$.

In Exercises 16–18, eliminate the parameter and write the corresponding rectangular equation.

16. $x = \sqrt{t + 1},\ \ y = 3 + t$

17. $x = 3 - 2\sin\theta,\ \ y = 1 + 5\cos\theta$

18. $x = e^{2t},\ \ y = e^{4t}$

19. Find a set of parametric equations for the circle with center $(5,\ -4)$ and radius 6.

20. Find a set of parametric equations for the hyperbola with vertices $(\pm 3,\ 0)$ and foci $(\pm 5,\ 0)$.

CHAPTER 1

Practice Test Solutions

1. $-|-17| - 17 = -17 - 17 = -34$

2. $d = \left|\frac{12}{5} - \left(-\frac{7}{15}\right)\right| = \left|\frac{36}{15} + \frac{7}{15}\right| = \frac{43}{15}$

3. $|z - (-5)| \leq 12$

$\quad |z + 5| \leq 12$

4. The midpoint of the interval is -2 and the distance between -2 and either endpoint is 6. Therefore, we have

$$|x - (-2)| \leq 6$$
$$|x + 2| \leq 6$$

5. $\dfrac{6x + 5}{2x - 9} = \dfrac{3}{5}$

$\quad 5(6x + 5) = 3(2x - 9)$

$\quad 30x + 25 = 6x - 27$

$\quad\quad\quad 24x = -52$

$\quad\quad\quad\quad x = -\dfrac{52}{24} = -\dfrac{13}{6}$

6. $4x^2 + 3x - 5 = 0$

$\quad a = 4,\ b = 3,\ c = -5$

$\quad x = \dfrac{-3 \pm \sqrt{(3)^2 - 4(4)(-5)}}{2(4)} = \dfrac{-3 \pm \sqrt{89}}{8}$

7. $21.4x^2 + 6.9x - 1.4 = 0$

$\quad a = 21.4,\ b = 6.9,\ c = -1.4$

$\quad x = \dfrac{-6.9 \pm \sqrt{(6.9)^2 - 4(21.4)(-1.4)}}{2(21.4)} = \dfrac{-6.9 \pm \sqrt{167.45}}{42.8}$

$\quad x = \dfrac{-6.9 + \sqrt{167.45}}{42.8} \approx 0.141$

$\quad x = \dfrac{-6.9 - \sqrt{167.48}}{42.8} \approx -0.464$

8. $x^6 - 7x^3 - 8 = 0$

$(x^3 - 8)(x^3 + 1) = 0$

$x^3 = 8$ OR $x^3 = -1$

$x = 2$ $x = -1$

9. $d = \sqrt{(-2-4)^2 + (5-7)^2} = \sqrt{36+4} = \sqrt{40} = 2\sqrt{10}$

10. $\left(\dfrac{-1+3}{2}, \dfrac{16+(-5)}{2}\right) = \left(1, \dfrac{11}{2}\right)$

11. $\sqrt{(-6-2)^2 + (x-0)^2} = 9$

$\sqrt{64 + x^2} = 9$

$64 + x^2 = 81$

$x^2 = 17$

$x = \pm\sqrt{17}$

12. $y = x\sqrt{3-x}$

x-intercepts: $0 = x\sqrt{3-x} \Rightarrow x = 0$ or $x = 3$

$(0,\ 0)$ and $(3,\ 0)$

y-intercept: $y = 0\sqrt{3-0} = 0$

$(0,\ 0)$

13. $y = \dfrac{(-x)^2}{(-x)^3 - 1} = \dfrac{x^2}{-x^3 - 1} \neq \dfrac{x^2}{x^3 - 1}$; *Not* symmetric with respect to the y-axis

$-y = \dfrac{x^2}{x^3 - 1} \Rightarrow y = -\dfrac{x^2}{x^3 - 1} \neq \dfrac{x^2}{x^3 - 1}$; *Not* symmetric with respect to the x-axis

$-y = \dfrac{(-x)^2}{(-x)^3 - 1} \Rightarrow -y = \dfrac{x^2}{-x^3 - 1} \Rightarrow y = \dfrac{x^2}{x^3 + 1} \neq \dfrac{x^2}{x^3 - 1}$; *Not* symmetric with respect to the origin

14. $y = \sqrt{x+2}$

x	-2	-1	2	7
y	0	1	2	3

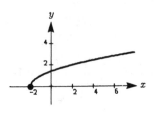

15. $y = |x - 3|$

x	0	1	2	3	4	5
y	3	2	1	0	1	2

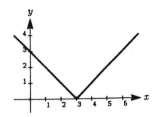

16. $\quad x^2 + y^2 - 14x + 6y + 42 = 0$

$$x^2 - 14x + 49 + y^2 + 6y + 9 = -42 + 49 + 9$$

$$(x - 7)^2 + (y + 3)^2 = 16$$

17. $\quad f(x) = 5x + 11$

$$f(2) = 10 + 11 = 21$$

$$\frac{f(x) - f(2)}{x - 2} = \frac{(5x + 11) - 21}{x - 2} = \frac{5x - 10}{x - 2} = \frac{5(x - 2)}{x - 2} = 5$$

18. $f(x) = \sqrt{\dfrac{x - 1}{x + 3}}$

Domain: $\dfrac{x - 1}{x + 3} \geq 0$

Zeros: $x = 1, \; x = -3$

$$\frac{(-)}{(-)} > 0 \quad \Big| \quad \frac{(-)}{(+)} < 0 \quad \Big| \quad \frac{(+)}{(+)} > 0$$

$$\quad \text{YES} \qquad\qquad \text{NO} \qquad\quad \text{YES}$$

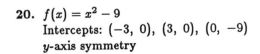

$x < -3$ OR $x \geq 1$

19. $V = l \cdot w \cdot h$

$$V = (16 - 2x)(11 - 2x)(x)$$

$$V = 4x^3 - 54x^2 + 176x$$

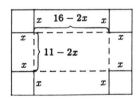

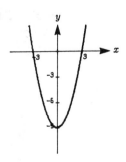

20. $f(x) = x^2 - 9$

Intercepts: $(-3, 0), \; (3, 0), \; (0, -9)$

y-axis symmetry

21. $f(x) = -1 + |x|$
Intercepts: $(-1, 0)$, $(1, 0)$, $(0, -1)$
y-axis symmetry

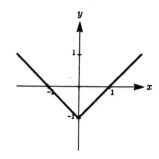

22.
$$f(x) = \begin{cases} 3x + 2, & x \geq 1 \\ x^2 - 1, & x < 1 \end{cases}$$

23. $f(x) = x^2 + 2; \quad g(x) = 3x - 8$
$$(f \circ g)(x) = f(g(x))$$
$$= f(3x - 8)$$
$$= (3x - 8)^2 + 2$$
$$= 9x^2 - 48x + 66$$

24. $\quad f(x) = \dfrac{x + 3}{x}$
$$y = \dfrac{x + 3}{x}$$
$$x = \dfrac{y + 3}{y}$$
$$xy = y + 3$$
$$xy - y = 3$$
$$y(x - 1) = 3$$
$$y = \dfrac{3}{x - 1}$$
$$f^{-1}(x) = \dfrac{3}{x - 1}$$

25. $\quad f(x) = \dfrac{x^3}{4}; \qquad g(x) = 3x$
$$f^{-1}(x) = \sqrt[3]{4x}; \quad g^{-1}(x) = \dfrac{x}{3}$$
$$g^{-1} \circ f^{-1} = g^{-1}(f^{-1}(x))$$
$$= g^{-1}(\sqrt[3]{4x})$$
$$= \dfrac{\sqrt[3]{4x}}{3}$$

CHAPTER 2

Practice Test Solutions

1. (a) $350° = 350\left(\dfrac{\pi}{180}\right) = \dfrac{35\pi}{18}$

(b) $\dfrac{5\pi}{9} = \dfrac{5\pi}{9} \cdot \dfrac{180}{\pi} = 100°$

2. (a) $135°14'12'' = 135 + \dfrac{14}{60} + \dfrac{12}{3600}$

$\approx 135.2367°$

(b) $-22.569° = -(22 + 0.569(60)')$

$= -22°34.14'$

$= -(22°34' + 0.14(60)'')$

$\approx -22°34'8''$

3. (a) $\dfrac{5\pi}{6}$ corresponds to the point $\left(-\dfrac{\sqrt{3}}{2}, \dfrac{1}{2}\right)$

$\sin\dfrac{5\pi}{6} = y = \dfrac{1}{2}$

(b) $\dfrac{5\pi}{4}$ corresponds to the point $\left(-\dfrac{\sqrt{2}}{2}, -\dfrac{\sqrt{2}}{2}\right)$

$\tan\dfrac{5\pi}{4} = \dfrac{y}{x} = 1$

4. (a) $\sin 7\pi = \sin(6\pi + \pi) = \sin\pi = 0$

(b) $\cos\left(-\dfrac{13\pi}{3}\right) = \cos\left(-4\pi - \dfrac{\pi}{3}\right)$

$= \cos\left(-\dfrac{\pi}{3}\right)$

$= \cos\dfrac{\pi}{3} = \dfrac{1}{2}$

5. $\cos\theta = \dfrac{2}{3}$

$x = 2,\ r = 3,\ y = \sqrt{9-4} = \sqrt{5}$

$\tan\theta = \dfrac{y}{x} = \dfrac{\sqrt{5}}{2}$

6. $\sin\theta = 0.9063$

$\theta = \arcsin(0.9063)$

$\theta \approx 65°$ or $\dfrac{13\pi}{36}$

7. $\tan 20° = \dfrac{35}{x}$

$x = \dfrac{35}{\tan 20°} \approx 96.1617$

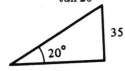

8. $\theta = \dfrac{6\pi}{5}$, θ is in Quadrant III.

Reference angle: $\dfrac{6\pi}{5} - \pi = \dfrac{\pi}{5}$ or $36°$

9. $\csc 3.92 = \dfrac{1}{\sin 3.92} \approx -1.4242$

10. $\tan\theta = 6 = \dfrac{6}{1}$, θ lies in Quadrant III.

$y = -6,\ x = -1,\ r = \sqrt{36+1} = \sqrt{37}$,

so $\sec\theta = \dfrac{\sqrt{37}}{-1} \approx -6.0828.$

11. Period: 4π

Amplitude: 3

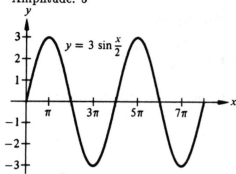

$$y = 3 \sin \frac{x}{2}$$

12. Period: 2π

Amplitude: 2

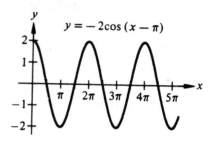

$$y = -2\cos(x - \pi)$$

13. Period: $\dfrac{\pi}{2}$

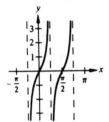

14. Period: 2π

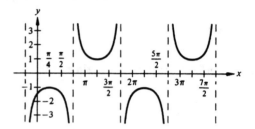

15.

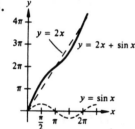

16.

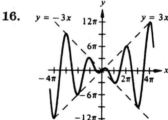

17. Let $\theta = \arcsin 1$

$$\sin \theta = 1$$

$$\theta = \frac{\pi}{2}$$

18. $\arctan(-3) = -\arctan 3$

$$\tan \theta = -3$$

$$\theta \approx -1.249 \text{ or } -71.565°$$

19. $\sin\left(\arccos \dfrac{4}{\sqrt{35}}\right)$

$$\sin \theta = \frac{\sqrt{19}}{\sqrt{35}} \approx 0.7368$$

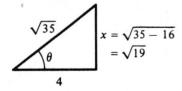

$$x = \sqrt{35 - 16}$$
$$= \sqrt{19}$$

20. $\cos\left(\arcsin \dfrac{x}{4}\right)$

$$\cos \theta = \frac{\sqrt{16 - x^2}}{4}$$

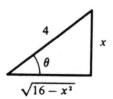

21. Given $A = 40°$, $c = 12$

$B = 90° - 40° = 50°$

$\sin 40° = \dfrac{a}{12}$

$a = 12 \sin 40° \approx 7.713$

$\cos 40° = \dfrac{b}{12}$

$b = 12 \cos 40° \approx 9.192$

22. Given $B = 6.84°$, $a = 21.3$

$A = 90° - 6.84° = 83.16°$

$\sin 83.16° = \dfrac{21.3}{c}$

$c = \dfrac{21.3}{\sin 83.16°} \approx 21.453$

$\tan 83.16° = \dfrac{21.3}{b}$

$b = \dfrac{21.3}{\tan 83.16°} \approx 2.555$

23. Given $a = 5$, $b = 9$

$c = \sqrt{25 + 81} = \sqrt{106} \approx 10.296$

$\tan A = \frac{5}{9}$

$A = \arctan \frac{5}{9} \approx 29.055°$

$B = 90° - 29.055° = 60.945°$

24. $\sin 67° = \dfrac{x}{20}$

$x = 20 \sin 67° \approx 18.41'$

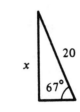

25. $\tan 5° = \dfrac{250}{x}$

$x = \dfrac{250}{\tan 5°}$

$\approx 2857.513'$

≈ 0.541 mi

CHAPTER 3

Practice Test Solutions

1. $\tan x = \dfrac{4}{11}$, $\sec x < 0 \Rightarrow x$ is in Quadrant III.

$y = -4$, $x = -11$, $r = \sqrt{16 + 121} = \sqrt{137}$

$$\sin x = -\frac{4}{\sqrt{137}} = -\frac{4\sqrt{137}}{137} \qquad \csc x = -\frac{\sqrt{137}}{4}$$

$$\cos x = -\frac{11}{\sqrt{137}} = -\frac{11\sqrt{137}}{137} \qquad \sec x = -\frac{\sqrt{137}}{11}$$

$$\tan x = \frac{4}{11} \qquad \cot x = \frac{11}{4}$$

2. $\dfrac{\sec^2 x + \csc^2 x}{\csc^2 x (1 + \tan^2 x)} = \dfrac{\sec^2 x + \csc^2 x}{\csc^2 x + (\csc^2 x)\tan^2 x} = \dfrac{\sec^2 x + \csc^2 x}{\csc^2 x + \dfrac{1}{\sin^2 x} \cdot \dfrac{\sin^2 x}{\cos^2 x}}$

$$= \frac{\sec^2 x + \csc^2 x}{\csc^2 x + \dfrac{1}{\cos^2 x}} = \frac{\sec^2 x + \csc^2 x}{\csc^2 x + \sec^2 x} = 1$$

3. $\ln|\tan\theta| - \ln|\cot\theta| = \ln\dfrac{|\tan\theta|}{|\cot\theta|} = \ln\left|\dfrac{\sin\theta/\cos\theta}{\cos\theta/\sin\theta}\right| = \ln\left|\dfrac{\sin^2\theta}{\cos^2\theta}\right| = \ln|\tan^2\theta| = 2\ln|\tan\theta|$

4. $\cos\left(\dfrac{\pi}{2} - x\right) = \dfrac{1}{\csc x}$ is true since $\cos\left(\dfrac{\pi}{2} - x\right) = \sin x = \dfrac{1}{\csc x}$.

5. $\sin^4 x + (\sin^2 x)\cos^2 x = \sin^2 x(\sin^2 x + \cos^2 x) = \sin^2 x(1) = \sin^2 x$

6. $(\csc x + 1)(\csc x - 1) = \csc^2 x - 1 = \cot^2 x$

7. $\dfrac{\cos^2 x}{1 - \sin x} \cdot \dfrac{1 + \sin x}{1 + \sin x} = \dfrac{\cos^2 x(1 + \sin x)}{1 - \sin^2 x} = \dfrac{\cos^2 x(1 + \sin x)}{\cos^2 x} = 1 + \sin x$

8. $\dfrac{1 + \cos\theta}{\sin\theta} + \dfrac{\sin\theta}{1 + \cos\theta} = \dfrac{(1 + \cos\theta)^2 + \sin^2\theta}{\sin\theta(1 + \cos\theta)}$

$$= \frac{1 + 2\cos\theta + \cos^2\theta + \sin^2\theta}{\sin\theta(1 + \cos\theta)} = \frac{2 + 2\cos\theta}{\sin\theta(1 + \cos\theta)} = \frac{2}{\sin\theta} = 2\csc\theta$$

9. $\tan^4 x + 2\tan^2 x + 1 = (\tan^2 x + 1)^2 = (\sec^2 x)^2 = \sec^4 x$

10. (a) $\sin 105° = \sin(60° + 45°) = \sin 60° \cos 45° + \cos 60° \sin 45°$

$$= \frac{\sqrt{3}}{2} \cdot \frac{\sqrt{2}}{2} + \frac{1}{2} \cdot \frac{\sqrt{2}}{2} = \frac{\sqrt{2}}{4}(\sqrt{3} + 1)$$

(b) $\tan 15° = \tan(60° - 45°) = \dfrac{\tan 60° - \tan 45°}{1 + \tan 60° \tan 45°}$

$$= \frac{\sqrt{3} - 1}{1 + \sqrt{3}} \cdot \frac{1 - \sqrt{3}}{1 - \sqrt{3}} = \frac{2\sqrt{3} - 1 - 3}{1 - 3} = \frac{2\sqrt{3} - 4}{-2} = 2 - \sqrt{3}$$

11. $(\sin 42°) \cos 38° - (\cos 42°) \sin 38° = \sin(42° - 38°) = \sin 4°$

12. $\tan\left(\theta + \dfrac{\pi}{4}\right) = \dfrac{\tan\theta + \tan(\pi/4)}{1 - (\tan\theta)\tan(\pi/4)} = \dfrac{\tan\theta + 1}{1 - \tan\theta(1)} = \dfrac{1 + \tan\theta}{1 - \tan\theta}$

13. $\sin(\arcsin x - \arccos x) = \sin(\arcsin x)\cos(\arccos x) - \cos(\arcsin x)\sin(\arccos x)$

$$= (x)(x) - (\sqrt{1 - x^2})(\sqrt{1 - x^2}) = x^2 - (1 - x^2) = 2x^2 - 1$$

14. (a) $\cos(120°) = \cos[2(60°)] = 2\cos^2 60° - 1 = 2\left(\dfrac{1}{2}\right)^2 - 1 = -\dfrac{1}{2}$

(b) $\tan(300°) = \tan[2(150°)] = \dfrac{2\tan 150°}{1 - \tan^2 150°} = \dfrac{2\sqrt{3}/3}{1 - (1/3)} = \sqrt{3}$

15. (a) $\sin 22.5° = \sin\dfrac{45°}{2} = \sqrt{\dfrac{1 - \cos 45°}{2}} = \sqrt{\dfrac{1 - \sqrt{2}/2}{2}} = \dfrac{\sqrt{2 - \sqrt{2}}}{2}$

(b) $\tan\dfrac{\pi}{12} = \tan\dfrac{\pi/6}{2} = \dfrac{\sin(\pi/6)}{1 + \cos(\pi/6)} = \dfrac{1/2}{1 + \sqrt{3}/2} = \dfrac{1}{2 + \sqrt{3}} = 2 - \sqrt{3}$

16. $\sin\theta = \dfrac{4}{5}$, θ lies in Quadrant II $\Rightarrow \cos\theta = -\dfrac{3}{5}$.

$$\cos\frac{\theta}{2} = \sqrt{\frac{1 + \cos\theta}{2}} = \sqrt{\frac{1 - 3/5}{2}} = \sqrt{\frac{2}{10}} = \frac{1}{\sqrt{5}} = \frac{\sqrt{5}}{5}$$

17. $(\sin^2 x)\cos^2 x = \dfrac{1 - \cos 2x}{2} \cdot \dfrac{1 + \cos 2x}{2} = \dfrac{1}{4}[1 - \cos^2 2x] = \dfrac{1}{4}\left[1 - \dfrac{1 + \cos 4x}{2}\right]$

$$= \frac{1}{8}[2 - (1 + \cos 4x)] = \frac{1}{8}[1 - \cos 4x]$$

18. $6(\sin 5\theta)\cos 2\theta = 6\left\{\dfrac{1}{2}[\sin(5\theta + 2\theta) + \sin(5\theta - 2\theta)]\right\} = 3[\sin 7\theta + \sin 3\theta]$

19. $\sin(x + \pi) + \sin(x - \pi) = 2\left(\sin\dfrac{[(x + \pi) + (x - \pi)]}{2}\right)\cos\dfrac{[(x + \pi) - (x - \pi)]}{2}$

$$= 2(\sin x)\cos\pi = -2\sin x$$

20. $\dfrac{\sin 9x + \sin 5x}{\cos 9x - \cos 5x} = \dfrac{2\sin 7x \cos 2x}{-2\sin 7x \sin 2x} = -\dfrac{\cos 2x}{\sin 2x} = -\cot 2x$

21. $\frac{1}{2}[\sin(u+v) - \sin(u-v)] = \frac{1}{2}\{(\sin u)\cos v + (\cos u)\sin v - [(\sin u)\cos v - (\cos u)\sin v]\}$
$$= \frac{1}{2}[2(\cos u)\sin v] = (\cos u)\sin v$$

22. $4\sin^2 x = 1$

$\qquad \sin^2 x = \dfrac{1}{4}$

$\qquad \sin x = \pm\dfrac{1}{2}$

$\sin x = \dfrac{1}{2} \qquad$ or $\qquad \sin x = -\dfrac{1}{2}$

$x = \dfrac{\pi}{6}$ or $\dfrac{5\pi}{6} \qquad\qquad x = \dfrac{7\pi}{6}$ or $\dfrac{11\pi}{6}$

23. $\tan^2\theta + (\sqrt{3}-1)\tan\theta - \sqrt{3} = 0$

$\qquad (\tan\theta - 1)(\tan\theta + \sqrt{3}) = 0$

$\tan\theta = 1 \qquad$ or $\qquad \tan\theta = -\sqrt{3}$

$\theta = \dfrac{\pi}{4}$ or $\dfrac{5\pi}{4} \qquad\quad \theta = \dfrac{2\pi}{3}$ or $\dfrac{5\pi}{3}$

24. $\qquad\qquad \sin 2x = \cos x$

$\qquad 2(\sin x)\cos x - \cos x = 0$

$\qquad\qquad \cos x(2\sin x - 1) = 0$

$\cos x = 0 \qquad$ or $\qquad \sin x = \dfrac{1}{2}$

$x = \dfrac{\pi}{2}$ or $\dfrac{3\pi}{2} \qquad\qquad x = \dfrac{\pi}{6}$ or $\dfrac{5\pi}{6}$

25. $\tan^2 x - 6\tan x + 4 = 0$

$$\tan x = \frac{-(-6) \pm \sqrt{(-6)^2 - 4(1)(4)}}{2(1)}$$

$$\tan x = \frac{6 \pm \sqrt{20}}{2} = 3 \pm \sqrt{5}$$

$\tan x = 3 + \sqrt{5} \qquad$ or $\qquad \tan x = 3 - \sqrt{5}$

$x = 1.3821$ or $4.5237 \qquad\qquad x = 0.6524$ or 3.7940

CHAPTER 4

Practice Test Solutions

1. $C = 180° - (40° + 12°) = 128°$

$a = \sin 40° \left(\dfrac{100}{\sin 12°} \right) \approx 309.164$

$c = \sin 128° \left(\dfrac{100}{\sin 12°} \right) \approx 379.012$

2. $\sin A = 5 \left(\dfrac{\sin 150°}{20} \right) = 0.1250$

$A \approx 7.181°$

$B \approx 180° - (150° + 7.181°) = 22.819°$

$b = \sin 22.819° \left(\dfrac{20}{\sin 150°} \right) \approx 15.513$

3. Area $= \frac{1}{2} ab \sin C$

$= \frac{1}{2}(3)(5) \sin 130°$

≈ 5.745 square units

4. $h = b \sin A, \ a = 10$

$= 35 \sin 22.5$

≈ 13.394

Since $a < h$ and A is acute, the triangle has no solution.

5. $\cos A = \dfrac{(53)^2 + (38)^2 - (49)^2}{2(53)(38)} \approx 0.4598$

$A \approx 62.627°$

$\cos B = \dfrac{(49)^2 + (38)^2 - (53)^2}{2(49)(38)} \approx 0.2782$

$B \approx 73.847°$

$C = 180° - (62.627° + 73.847°) = 43.526°$

6. $c^2 = (100)^2 + (300)^2 - 2(100)(300) \cos 29°$

≈ 47522.8176

$c \approx 218$

$\cos A = \dfrac{(300)^2 + (218)^2 - (100)^2}{2(300)(218)} \approx 0.9750$

$A \approx 12.85°$

$B = 180° - (12.85° + 29°) = 138.15°$

7. $s = \dfrac{a+b+c}{2} = \dfrac{4.1 + 6.8 + 5.5}{2} = 8.2$

Area $= \sqrt{s(s-a)(s-b)(s-c)}$

$= \sqrt{8.2(8.2 - 4.1)(8.2 - 6.8)(8.2 - 5.5)}$

≈ 11.273 square units

8. $x^2 = (40)^2 + (70)^2 - 2(40)(70) \cos 168°$

≈ 11977.6266

$x \approx 109.442$ miles

9. $\mathbf{w} = 4(3\mathbf{i} + \mathbf{j}) - 7(-\mathbf{i} + 2\mathbf{j})$

$= 19\mathbf{i} - 10\mathbf{j}$

10. $\dfrac{\mathbf{v}}{\|\mathbf{v}\|} = \dfrac{5\mathbf{i} - 3\mathbf{j}}{\sqrt{25 + 9}} = \dfrac{5}{\sqrt{34}}\mathbf{i} - \dfrac{3}{\sqrt{34}}\mathbf{j}$

$= \dfrac{5\sqrt{34}}{34}\mathbf{i} - \dfrac{3\sqrt{34}}{34}\mathbf{j}$

11. $u = 6i + 5j \qquad v = 2i - 3j \qquad w = -4i - 8j$

$\|u\| = \sqrt{61} \qquad \|v\| = \sqrt{13} \qquad \|w\| = \sqrt{80}$

$\cos\theta = \dfrac{61 + 13 - 80}{2\sqrt{61}\sqrt{13}}$

$\theta \approx 96.116°$

12. $\tan 30° = \dfrac{y}{x} = \dfrac{1}{\sqrt{3}}$

$u = \sqrt{3}\,i + j \text{ but } \|u\| = 2$

$v = 2u = 2\sqrt{3}\,i + 2j$

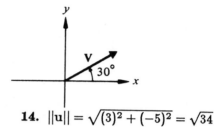

13. $u \cdot v = 3(-2) + (-5)(1) = -11$

14. $\|u\| = \sqrt{(3)^2 + (-5)^2} = \sqrt{34}$

15. $\cos\theta = \dfrac{-11}{\sqrt{34}\sqrt{5}}$

$\theta = \arccos\left(\dfrac{-11}{\sqrt{170}}\right) \approx 147.5288°$

16. $(v \cdot u)u = (u \cdot v)u = -11u = \langle -33,\ 55 \rangle$

17. $u \cdot v = -11 \neq 0 \Rightarrow$ *not* orthogonal

$u \neq kv$ for any real number $k \Rightarrow$ *not* parallel

Neither

18. $\text{proj}_v(u) = \left(\dfrac{-11}{5}\right)v = \left\langle \dfrac{22}{5},\ -\dfrac{11}{5} \right\rangle$

19. $u - \text{proj}_v(u) = \langle 3,\ -5 \rangle - \left\langle \dfrac{22}{5},\ -\dfrac{11}{5} \right\rangle = \left\langle -\dfrac{7}{5},\ -\dfrac{14}{5} \right\rangle$

20. $W = (\cos 40°)(70)(20) = 1072.4622$ ft-lb

CHAPTER 5

Practice Test Solutions

1. $i^{38} = i^{36}i^2 = (i^4)^9 i^2 = (1)^9(-1) = -1$

2. $(8 + \sqrt{-64}) + (6 + \sqrt{-25}) = (8 + 8i) + (6 + 5i) = 14 + 13i$

3. $-(4 + 4i) - (-3i) = -4 - 4i + 3i = -4 - i$

4. $(-8 + 2i)(-8 - 2i) = 64 + 16i - 16i - 4i^2 = 64 + 4 = 68 + 0i$

5.
$$\frac{12 + 16i}{4 - 2i} = \frac{6 + 8i}{2 - i} \cdot \frac{2 + i}{2 + i}$$
$$= \frac{12 + 6i + 16i + 8i^2}{4 + 1} = \frac{4 + 22i}{5}$$
$$= \frac{4}{5} + \frac{22}{5}i$$

6. $3x^2 + 2x + 2 = 0$
$$x = \frac{-2 \pm \sqrt{(2)^2 - 4(3)(2)}}{2(3)} = \frac{-2 \pm \sqrt{-20}}{6}$$
$$= \frac{-2 \pm 2i\sqrt{5}}{6} = -\frac{1}{3} \pm \frac{\sqrt{5}}{3}i$$

7. $3x^2 + 1 = -47$
$$3x^2 = -48$$
$$x^2 = -16$$
$$x = \pm\sqrt{-16}$$
$$x = \pm 4i = 0 \pm 4i$$

8.
$$x^4 - 1296 = 0$$
$$(x^2 + 36)(x^2 - 36) = 0$$
$$x^2 = -36 \quad \text{OR} \quad x^2 = 36$$
$$x = \pm\sqrt{-36} \qquad\qquad x = \pm\sqrt{36}$$
$$x = \pm 6i \qquad\qquad\quad x = \pm 6$$
$$x = 0 \pm 6i$$

9.
$$x^4 - 7x^2 - 60 = 0$$
$$(x^2 - 12)(x^2 + 5) = 0$$
$$x^2 = 12 \quad \text{OR} \quad x^2 = -5$$
$$x = \pm\sqrt{12} \qquad\qquad x = \pm\sqrt{-5}$$
$$x = \pm 2\sqrt{3} \qquad\quad x = \pm\sqrt{5}\,i = 0 \pm \sqrt{5}\,i$$

10.
$$x^3 + 2x^2 + 9x + 18 = 0$$
$$x^2(x + 2) + 9(x + 2) = 0$$
$$(x + 2)(x^2 + 9) = 0$$
$$x = -2 \quad \text{OR} \quad x^2 = -9$$
$$x = \pm 3i = 0 \pm 3i$$

11. Since $-2i$ is a zero, so is $2i$.
$$P(x) = (x - 0)(x + 2i)(x - 2i)(x - 3)$$
$$= x(x^2 + 4)(x - 3)$$
$$= x(x^3 - 3x^2 + 4x - 12)$$
$$= x^4 - 3x^3 + 4x^2 - 12x$$

12. Since $1 + 2i$ is a zero, so is $1 - 2i$. Thus,

$$[x - (1 + 2i)][x - (1 - 2i)] = [(x - 1) - 2i][(x - 1) + 2i]$$
$$= (x - 1)^2 - 4i^2$$
$$= x^2 - 2x + 5$$

is a factor of $f(x)$.

$$
\begin{array}{r}
x\ +6 \\
x^2 - 2x + 5\overline{\smash{\big)}\ x^3 + 4x^2 -\ \ 7x + 30} \\
\underline{x^3 - 2x^2 +\ \ 5x} \\
6x^2 - 12x + 30 \\
\underline{6x^2 - 12x + 30} \\
0
\end{array}
$$

Therefore, $f(x) = (x^2 - 2x + 5)(x + 6)$ and the corresponding zeros are -6, $1 + 2i$, $1 - 2i$.

13. $r = \sqrt{25 + 25} = \sqrt{50} = 5\sqrt{2}$

$\tan \theta = \dfrac{-5}{5} = -1$

Since z is in Quadrant IV,

$\theta = 315°$

$z = 5\sqrt{2}(\cos 315° + i \sin 315°)$.

14. $\cos 225° = -\dfrac{\sqrt{2}}{2}$ $\quad \sin 225° = -\dfrac{\sqrt{2}}{2}$

$z = 6\left(-\dfrac{\sqrt{2}}{2} - i\dfrac{\sqrt{2}}{2}\right)$

$ = -3\sqrt{2} - 3\sqrt{2}\,i$

15. $[7(\cos 23° + i \sin 23°)][4(\cos 7° + i \sin 7°)] = 7(4)[\cos(23° + 7°) + i \sin(23° + 7°)]$
$$= 28(\cos 30° + i \sin 30°)$$

16. $\dfrac{9\left(\cos \dfrac{5\pi}{4} + i \sin \dfrac{5\pi}{4}\right)}{3(\cos \pi + i \sin \pi)} = \dfrac{9}{3}\left[\cos\left(\dfrac{5\pi}{4} - \pi\right) + i \sin\left(\dfrac{5\pi}{4} - \pi\right)\right] = 3\left(\cos \dfrac{\pi}{4} + i \sin \dfrac{\pi}{4}\right)$

17. $(2 + 2i)^8 = [2\sqrt{2}(\cos 45° + i \sin 45°)]^8 = (2\sqrt{2})^8[\cos(8)(45°) + i \sin(8)(45°)]$
$$= 4096[\cos 360° + i \sin 360°] = 4096$$

18. $z = 8\left(\cos \dfrac{\pi}{3} + i \sin \dfrac{\pi}{3}\right)$, $\quad n = 3$; The cube roots of z are:

For $K = 0$, $\sqrt[3]{8}\left[\cos \dfrac{\pi/3}{3} + i \sin \dfrac{\pi/3}{3}\right] = 2\left(\cos \dfrac{\pi}{9} + i \sin \dfrac{\pi}{9}\right)$

For $K = 1$, $\sqrt[3]{8}\left[\cos \dfrac{\pi/3 + 2\pi}{3} + i \sin \dfrac{\pi/3 + 2\pi}{3}\right] = 2\left(\cos \dfrac{7\pi}{9} + i \sin \dfrac{7\pi}{9}\right)$

For $K = 2$, $\sqrt[3]{8}\left[\cos \dfrac{\pi/3 + 4\pi}{3} + i \sin \dfrac{\pi/3 + 4\pi}{3}\right] = 2\left(\cos \dfrac{13\pi}{9} + i \sin \dfrac{13\pi}{9}\right)$

19. $x^3 = -125 = 125(\cos \pi + i \sin \pi)$

For $K = 0$, $\sqrt[3]{125}\left(\cos \dfrac{\pi}{3} + i \sin \dfrac{\pi}{3}\right) = 5\left(\cos \dfrac{\pi}{3} + i \sin \dfrac{\pi}{3}\right)$

For $K = 1$, $\sqrt[3]{125}\left(\cos \dfrac{\pi + 2\pi}{3} + i \sin \dfrac{\pi + 2\pi}{3}\right) = -5$

For $K = 2$, $\sqrt[3]{125}\left(\cos \dfrac{\pi + 4\pi}{3} + i \sin \dfrac{\pi + 4\pi}{3}\right) = 5\left(\cos \dfrac{5\pi}{3} + i \sin \dfrac{5\pi}{3}\right)$

20. $x^4 = -i = 1\left(\cos \dfrac{3\pi}{2} + i \sin \dfrac{3\pi}{2}\right)$

For $K = 0$, $\cos \dfrac{3\pi/2}{4} + i \sin \dfrac{3\pi/2}{4} = \cos \dfrac{3\pi}{8} + i \sin \dfrac{3\pi}{8}$

For $K = 1$, $\cos \dfrac{3\pi/2 + 2\pi}{4} + i \sin \dfrac{3\pi/2 + 2\pi}{4} = \cos \dfrac{7\pi}{8} + i \sin \dfrac{7\pi}{8}$

For $K = 2$, $\cos \dfrac{3\pi/2 + 4\pi}{4} + i \sin \dfrac{3\pi/2 + 4\pi}{4} = \cos \dfrac{11\pi}{8} + i \sin \dfrac{11\pi}{8}$

For $K = 3$, $\cos \dfrac{3\pi/2 + 6\pi}{4} + i \sin \dfrac{3\pi/2 + 6\pi}{4} = \cos \dfrac{15\pi}{8} + i \sin \dfrac{15\pi}{8}$

CHAPTER 6

Practice Test Solutions

1. $x^{3/5} = 8$

$x = 8^{5/3} = (\sqrt[3]{8})^5 = 2^5 = 32$

2. $3^{x-1} = \frac{1}{81}$

$3^{x-1} = 3^{-4}$

$x - 1 = -4$

$x = -3$

3. $f(x) = 2^{-x} = \left(\frac{1}{2}\right)^x$

x	-2	-1	0	1	2
$f(x)$	4	2	1	$\frac{1}{2}$	$\frac{1}{4}$

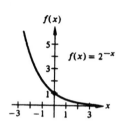

4. $g(x) = e^x + 1$

x	-2	-1	0	1	2
$g(x)$	1.14	1.37	2	3.72	8.39

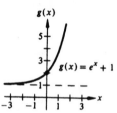

5. $A = P\left(1 + \dfrac{r}{n}\right)^{nt}$

(a) $A = 5000\left(1 + \dfrac{0.09}{12}\right)^{12(3)} \approx \$6543.23$

(b) $A = 5000\left(1 + \dfrac{0.09}{4}\right)^{4(3)} \approx \$6530.25$

(c) $A = 5000e^{(0.09)(3)} \approx \$6549.82$

6. $7^{-2} = \frac{1}{49}$

$\log_7 \frac{1}{49} = -2$

7. $x - 4 = \log_2 \frac{1}{64}$

$2^{x-4} = \frac{1}{64}$

$2^{x-4} = 2^{-6}$

$x - 4 = -6$

$x = -2$

8. $\log_b \sqrt[4]{8/25} = \frac{1}{4}\log_b \frac{8}{25}$

$= \frac{1}{4}[\log_b 8 - \log_b 25]$

$= \frac{1}{4}[\log_b 2^3 - \log_b 5^2]$

$= \frac{1}{4}[3\log_b 2 - 2\log_b 5]$

$= \frac{1}{4}[3(0.3562) - 2(0.8271)]$

$= -0.1464$

9. $5\ln x - \dfrac{1}{2}\ln y + 6\ln z = \ln x^5 - \ln\sqrt{y} + \ln z^6$

$= \ln\left(\dfrac{x^5 z^6}{\sqrt{y}}\right)$

10. $\log_9 28 = \dfrac{\log 28}{\log 9} \approx 1.5166$

11. $\log N = 0.6646$

$N = 10^{0.6646} \approx 4.62$

12.

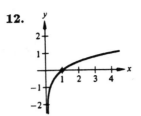

13. Domain: $x^2 - 9 > 0$

$(x + 3)(x - 3) > 0$

$x < -3 \text{ or } x > 3$

14.

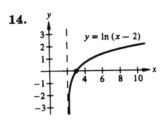

$y = \ln(x - 2)$

15. $\dfrac{\ln x}{\ln y} \neq \ln(x - y)$ since $\dfrac{\ln x}{\ln y} = \log_y x$

16. $5^x = 41$

$x = \log_5 41 = \dfrac{\ln 41}{\ln 5} \approx 2.3074$

17. $x - x^2 = \log_5 \frac{1}{25}$

$5^{x - x^2} = \frac{1}{25}$

$5^{x - x^2} = 5^{-2}$

$x - x^2 = -2$

$0 = x^2 - x - 2$

$0 = (x + 1)(x - 2)$

$x = -1 \text{ or } x = 2$

18. $\log_2 x + \log_2 (x - 3) = 2$

$\log_2 [x(x - 3)] = 2$

$x(x - 3) = 2^2$

$x^2 - 3x = 4$

$x^2 - 3x - 4 = 0$

$(x + 1)(x - 4) = 0$

$x = 4$

$x = -1 \text{ (extraneous solution)}$

19. $\dfrac{e^x + e^{-x}}{3} = 4$

$e^x(e^x + e^{-x}) = 12e^x$

$e^{2x} + 1 = 12e^x$

$e^{2x} - 12e^x + 1 = 0$

$e^x = \dfrac{12 \pm \sqrt{144 - 4}}{2}$

$e^x = 11.9161 \quad \text{or} \quad e^x = 0.0839$

$x = \ln 11.9161 \qquad x = \ln 0.0839$

$x \approx 2.4779 \qquad\quad x \approx -2.4779$

20. $A = Pe^{rt}$

$12,000 = 6,000e^{0.13t}$

$2 = e^{0.13t}$

$0.13t = \ln 2$

$t = \dfrac{\ln 2}{0.13}$

$t \approx 5.3319 \text{ yrs or 5 yrs 4 months}$

CHAPTER 7

Practice Test Solutions

1. $m = \dfrac{7-3}{-2-4} = \dfrac{4}{-6} = -\dfrac{2}{3}$

2. $3x - 10y + 40 = 0$

$$-10y = -3x - 40$$
$$y = \tfrac{3}{10}x + 4$$

Slope: $m = \tfrac{3}{10}$

y-intercept: $(0, 4)$

3. $m = \dfrac{15-6}{-7-(-8)} = 9$

$$y - 6 = 9[x - (-8)]$$
$$y - 6 = 9(x + 8)$$
$$y - 6 = 9x + 72$$
$$y = 9x + 78$$

4. $y - 5 = -\tfrac{1}{3}[x - (-1)]$

$$-3y + 15 = x + 1$$
$$0 = x + 3y - 14$$

5. $\dfrac{x}{6} + \dfrac{y}{-2} = 1$

$$x - 3y = 6$$

6. $W = 0.055S + 1950$

7. $x - 4y + 8 = 0 \Rightarrow y = \tfrac{1}{4}x + 2$

The slope of the given line is $m = \tfrac{1}{4}$. The slope of the parallel line must equal the slope of the given line.

$$y - (-7) = \tfrac{1}{4}(x - 4)$$
$$y + 7 = \tfrac{1}{4}x - 1$$
$$4y + 28 = x - 4$$
$$0 = x - 4y - 32$$

8. The line, $x = 6$, is vertical. The perpendicular line must be horizontal. The horizontal line through $(11, 8)$ is $y = 8$.

9. $(0, 2), (-1, -1): \quad m_1 = \dfrac{-1-2}{-1-0} = 3$

$(-1, -1), (5, 15): \quad m_2 = \dfrac{15-(-1)}{5-(-1)} = \dfrac{16}{6} = \dfrac{8}{3}$

Since $m_1 \neq m_2$, the points are *not* collinear.

10. $d = \dfrac{|2(9) + (-4)(4) + 7|}{\sqrt{(2)^2 + (-4)^2}} = \dfrac{9}{\sqrt{20}} = \dfrac{9}{2\sqrt{5}} = \dfrac{9\sqrt{5}}{10}$

11. $3x + 9y = 20 \Rightarrow y = -\frac{1}{3}x + \frac{20}{3}$

$\tan \theta = -\frac{1}{3}$

$\theta = \arctan\left(-\frac{1}{3}\right) \approx -18.4349°$

12. $y = 3x - 4 \Rightarrow m_1 = 3$

$y = -2x + 10 \Rightarrow m_2 = -2$

$\tan \theta = \dfrac{-2 - 3}{1 + (3)(-2)} = 1$

$\theta = \arctan 1 = 45°$

13. $x^2 - 6x - 4y + 1 = 0$

$x^2 - 6x + 9 = 4y - 1 + 9$

$(x - 3)^2 = 4y + 8$

$(x - 3)^2 = 4(1)(y + 2) \Rightarrow p = 1$

Vertex: $(3, -2)$

Focus: $(3, -1)$

Directrix: $y = -3$

14. Vertex: $(2, -5)$

Focus: $(2, -6)$

Vertical axis;

opens downward with $p = -1$

$(x - h)^2 = 4p(y - k)$

$(x - 2)^2 = 4(-1)(y + 5)$

$x^2 - 4x + 4 = -4y - 20$

$x^2 - 4x + 4y + 24 = 0$

15. $x^2 + 4y^2 - 2x + 32y + 61 = 0$

$(x^2 - 2x + 1) + 4(y^2 + 8y + 16) = -61 + 1 + 64$

$(x - 1)^2 + 4(y + 4)^2 = 4$

$\dfrac{(x - 1)^2}{4} + \dfrac{(y + 4)^2}{1} = 1$

$a = 2,\ b = 1,\ c = \sqrt{3}$

Horizontal major axis

Center: $(1, -4)$

Foci: $(1 \pm \sqrt{3}, -4)$

Vertices: $(3, -4),\ (-1, -4)$

Eccentricity: $e = \sqrt{3}/2$

16. Vertices: $(0, \pm 6)$

Eccentricity: $e = 1/2$

Center: $(0, 0)$

Vertical major axis

$a = 6,\ e = \dfrac{c}{a} = \dfrac{c}{6} = \dfrac{1}{2} \Rightarrow c = 3,$

$b^2 = (6)^2 - (3)^2 = 27$

$\dfrac{x^2}{27} + \dfrac{y^2}{36} = 1$

17.
$$16y^2 - x^2 - 6x - 128y + 231 = 0$$
$$16(y^2 - 8y + 16) - (x^2 + 6x + 9) = -231 + 256 - 9$$
$$16(y - 4)^2 - (x + 3)^2 = 16$$
$$\frac{(y - 4)^2}{1} - \frac{(x + 3)^2}{16} = 1$$

$a = 1, \; b = 4, \; c = \sqrt{17}$

Center: $(-3, \; 4)$; vertical transverse axis

Vertices: $(-3, \; 5), \; (-3, \; 3)$

Foci: $(-3, \; 4 \pm \sqrt{17})$

Asymptotes: $y = 4 \pm \frac{1}{4}(x + 3)$

18. Vertices: $(\pm 3, \; 2)$

Foci: $(\pm 5, \; 2)$

Center: $(0, \; 2)$; horizontal transverse axis

$a = 3, \; c = 5, \; b = 4$
$$\frac{(x - 0)^2}{9} - \frac{(y - 2)^2}{16} = 1$$
$$\frac{x^2}{9} - \frac{(y - 2)^2}{16} = 1$$

19. $5x^2 + 2xy + 5y^2 - 10 = 0$

$A = 5, \; B = 2, \; C = 5, \; D = 0, \; E = 0, \; F = -10$

$$\cot 2\theta = \frac{5 - 5}{2} = 0$$
$$2\theta = \frac{\pi}{2} \Rightarrow \theta = \frac{\pi}{4}$$

$A' = 5\cos^2 \frac{\pi}{4} + 2\cos \frac{\pi}{4} \sin \frac{\pi}{4} + 5\sin^2 \frac{\pi}{4} = 6$

$C' = 5\sin^2 \frac{\pi}{4} - 2\cos \frac{\pi}{4} \sin \frac{\pi}{4} + 5\cos^2 \frac{\pi}{4} = 4$

$D' = E' = 0$

$F' = -10$

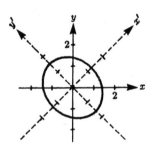

$$6(x')^2 + 4(y')^2 - 10 = 0$$
$$\frac{3(x')^2}{5} + \frac{2(y')^2}{5} = 1$$
$$\frac{(x')^2}{5/3} + \frac{(y')^2}{5/2} = 1$$

Ellipse centered at the origin

20. (a) $6x^2 - 2xy + y^2 = 0$

$A = 6, \ B = -2, \ C = 1$

$B^2 - 4AC = (-2)^2 - 4(6)(1) = -20 < 0$ Ellipse

(b) $x^2 + 4xy + 4y^2 - x - y + 17 = 0$

$A = 1, \ B = 4, \ C = 4$

$B^2 - 4AC = (4)^2 - 4(1)(4) = 0$ Parabola

CHAPTER 8

Practice Test Solutions

1. Polar: $\left(\sqrt{2}, \dfrac{3\pi}{4}\right)$

$$x = \sqrt{2}\cos\frac{3\pi}{4} = \sqrt{2}\left(-\frac{1}{\sqrt{2}}\right) = -1$$

$$y = \sqrt{2}\sin\frac{3\pi}{4} = \sqrt{2}\left(\frac{1}{\sqrt{2}}\right) = 1$$

Rectangular: $(-1,\ 1)$

2. Rectangular: $(\sqrt{3},\ -1)$

$$r = \pm\sqrt{(\sqrt{3})^2 + (-1)^2} = \pm 2$$

$$\tan\theta = \frac{\sqrt{3}}{-1} = -\sqrt{3}$$

$$\theta = \frac{2\pi}{3} \quad\text{or}\quad \theta = \frac{5\pi}{3}$$

Polar: $\left(-2,\ \dfrac{2\pi}{3}\right)$ or $\left(2,\ \dfrac{5\pi}{3}\right)$

3. Rectangular: $4x - 3y = 12$

Polar: $4r\cos\theta - 3r\sin\theta = 12$

$$r(4\cos\theta - 3\sin\theta) = 12$$

$$r = \frac{12}{4\cos\theta - 3\sin\theta}$$

4. Polar: $r = 5\cos\theta$

$$r^2 = 5r\cos\theta$$

Rectangular: $\quad x^2 + y^2 = 5x$

$$x^2 + y^2 - 5x = 0$$

5. Polar: $\quad r = \dfrac{3}{4 + \cos\theta}$

$$4r + r\cos\theta = 3$$

Rectangular: $\quad 4(\pm\sqrt{x^2 + y^2}) + x = 3$

$$4(\pm\sqrt{x^2 + y^2}) = 3 - x$$

$$16(x^2 + y^2) = (3 - x)^2$$

$$16x^2 + 16y^2 = 9 - 6x + x^2$$

$$15x^2 + 16y^2 + 6x - 9 = 0$$

6. $r = 1 - \cos\theta$

Cardioid

Symmetry: Polar axis

Maximum value of $|r|$: $r = 2$ when $\theta = \pi$

Zero of r: $r = 0$ when $\theta = 0$

θ	0	$\dfrac{\pi}{2}$	π	$\dfrac{3\pi}{2}$
r	0	1	2	1

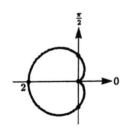

7. $\theta = \dfrac{3\pi}{4}$

Line

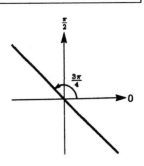

8. $r = 4\csc\theta$

$r = \dfrac{4}{\sin\theta}$

$r\sin\theta = 4$

$y = 4$

Horizontal line

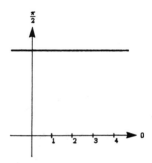

9. $r^2 = 9\cos 2\theta$

Lemniscate

Symmetry: Polar axis, $\theta = \pi/2$, and pole

Maximum value of $|r|$: $r = \pm 3$ when $\theta = 0$.

Zero of r : $r = 0$ when $\theta = \pi/4$.

θ	0	$\dfrac{\pi}{4}$	$\dfrac{\pi}{6}$
r	± 3	0	$\pm\dfrac{3}{\sqrt{2}}$

10. $r = 5\sin 2\theta$

Rose curve with four petals

Symmetry: Polar axis, $\theta = \dfrac{\pi}{2}$, and pole

Maximum value of $|r|$: $|r| = 5$ when $\theta = \dfrac{\pi}{4},\ \dfrac{3\pi}{4},\ \dfrac{5\pi}{4},\ \dfrac{7\pi}{4}$

Zeros of r: $r = 0$ when $\theta = 0,\ \dfrac{\pi}{2},\ \pi,\ \dfrac{3\pi}{2}$

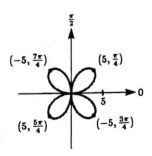

11.
$$r = \frac{12}{6\sin\theta - \cos\theta}$$
$$6r\sin\theta - r\cos\theta = 12$$
$$6y - x = 12$$
$$y = \frac{1}{6}x + 2$$

Line

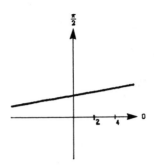

12. $r = \dfrac{3}{6 - \cos\theta}$

$$r = \frac{\frac{1}{2}}{1 - \frac{1}{6}\cos\theta}$$

$e = \frac{1}{6} < 1$, so the graph is an ellipse.

θ	0	$\dfrac{\pi}{2}$	π	$\dfrac{3\pi}{2}$
r	$\dfrac{3}{5}$	$\dfrac{1}{2}$	$\dfrac{3}{7}$	$\dfrac{1}{2}$

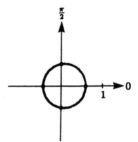

13. $r = \dfrac{5}{4 + 4\sin\theta}$

$$r = \frac{\frac{5}{4}}{1 + \sin\theta}$$

$e = 1$, so the graph is a parabola.

θ	0	$\dfrac{\pi}{2}$	π
r	$\dfrac{5}{4}$	$\dfrac{5}{8}$	$\dfrac{5}{4}$

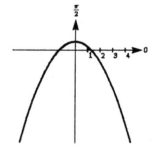

14. $r = \dfrac{10}{1 - 5\cos\theta}$

$e = 5 > 1$, so the graph is a hyperbola.

θ	0	$\dfrac{\pi}{2}$	π	$\dfrac{3\pi}{2}$
r	$-\dfrac{5}{2}$	10	$\dfrac{5}{3}$	10

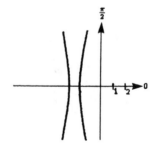

15. Parabola

Vertex: $\left(6, \dfrac{\pi}{2}\right)$

Focus: $(0, 0)$

$e = 1$

$r = \dfrac{ep}{1 + e\sin\theta}$

$r = \dfrac{p}{1 + \sin\theta}$

$6 = \dfrac{p}{1 + \sin(\pi/2)}$

$6 = \dfrac{p}{2}$

$12 = p$

$r = \dfrac{12}{1 + \sin\theta}$

16. $x = \sqrt{t + 1}, \quad y = 3 + t$

$x \geq 0, \quad t = x^2 - 1$

$y = 3 + (x^2 - 1)$

$y = 2 + x^2, \quad x \geq 0$

17. $x = 3 - 2\sin\theta, \quad y = 1 + 5\cos\theta$

$\dfrac{x - 3}{-2} = \sin\theta, \quad \dfrac{y - 1}{5} = \cos\theta$

$\left(\dfrac{x - 3}{-2}\right)^2 + \left(\dfrac{y - 1}{5}\right)^2 = 1$

$\dfrac{(x - 3)^2}{4} + \dfrac{(y - 1)^2}{25} = 1$

18. $x = e^{2t}, \quad y = e^{4t}$

$x > 0, \quad y > 0$

$x = e^{2t} \Rightarrow \ln x = 2t \Rightarrow t = \tfrac{1}{2}\ln x$

$y = e^{4t} = e^{4(1/2\ln x)} = e^{2\ln x} = e^{\ln x^2} = x^2$

$y = x^2, \quad x > 0, \quad y > 0$

Alternate solution:

$y = e^{4t} = \left(e^{2t}\right)^2 = x^2, \quad x > 0, \quad y > 0$

19. Center: $(5, -4)$

Radius: 6

$x = h + r\cos\theta, \quad y = k + r\sin\theta$

$x = 5 + 6\cos\theta, \quad y = -4 + 6\sin\theta$

20. Vertices: $(\pm3, 0)$

Foci: $(\pm5, 0)$

Center: $(0, 0)$

$a = 3, \ c = 5, \ b = 4$

$x = h + a\sec\theta, \quad y = k + b\tan\theta$

$x = 3\sec\theta, \qquad y = 4\tan\theta$